T0304742

# LECTURES ON $N_X(p)$

Research Notes in Mathematics

Volume 11

Editorial Board

Brian Conrey

Etienne Ghys

Bjorn Poonen

Peter Sarnak

Yuri Tschinkel

# LECTURES ON $N_X(p)$

JEAN-PIERRE SERRE

CRC Press is an imprint of the
Taylor & Francis Group, an **informa** business
AN A K PETERS BOOK

CRC Press
Taylor & Francis Group
6000 Broken Sound Parkway NW, Suite 300
Boca Raton, FL 33487-2742

© 2012 by Taylor & Francis Group, LLC
CRC Press is an imprint of Taylor & Francis Group, an Informa business

No claim to original U.S. Government works

Version Date: 20111005

International Standard Book Number: 978-1-4665-0192-8 (Hardback)

This book contains information obtained from authentic and highly regarded sources. Reasonable efforts have been made to publish reliable data and information, but the author and publisher cannot assume responsibility for the validity of all materials or the consequences of their use. The authors and publishers have attempted to trace the copyright holders of all material reproduced in this publication and apologize to copyright holders if permission to publish in this form has not been obtained. If any copyright material has not been acknowledged please write and let us know so we may rectify in any future reprint.

Except as permitted under U.S. Copyright Law, no part of this book may be reprinted, reproduced, transmitted, or utilized in any form by any electronic, mechanical, or other means, now known or hereafter invented, including photocopying, microfilming, and recording, or in any information storage or retrieval system, without written permission from the publishers.

For permission to photocopy or use material electronically from this work, please access www.copyright.com (http://www.copyright.com/) or contact the Copyright Clearance Center, Inc. (CCC), 222 Rosewood Drive, Danvers, MA 01923, 978-750-8400. CCC is a not-for-profit organization that provides licenses and registration for a variety of users. For organizations that have been granted a photocopy license by the CCC, a separate system of payment has been arranged.

**Trademark Notice:** Product or corporate names may be trademarks or registered trademarks, and are used only for identification and explanation without intent to infringe.

**Library of Congress Cataloging-in-Publication Data**

Serre, Jean-Pierre, 1926-
Lectures on N_X (p) / Jean-Pierre Serre.
p. cm. -- (Research notes in mathematics ; v. 11)
"An A K Peters book."
Summary: "This book deals with the question on how NX(p), the number of solutions of mod p congruences, varies with p when the family (X) of polynomial equations is fixed. While such a general question cannot have a complete answer, it offers a good occasion for reviewing various techniques in ℓ-adic cohomology and group representations, presented in a context that is appealing to specialists in number theory and algebraic geometry."-- Provided by publisher
Includes bibliographical references and indexes.
ISBN 978-1-4665-0192-8 (hardback)
1. Polynomials. 2. Number theory. 3. Representations of groups. 4. Cohomology operations. I. Title.

QA161.P59S44 2012
512.9'422--dc23 2011035770

**Visit the Taylor & Francis Web site at**
**http://www.taylorandfrancis.com**

**and the CRC Press Web site at**
**http://www.crcpress.com**

# Contents

Preface vii

Conventions ix

Chapter 1. Introduction 1
1.1. Definition of $N_X(p)$ : the affine case . . . . . . . . . . . . 1
1.2. Definition of $N_X(p)$ : the scheme setting . . . . . . . . . . 1
1.3. How large is $N_X(p)$ when $p \to \infty$? . . . . . . . . . . . . 2
1.4. More properties of $p \mapsto N_X(p)$ . . . . . . . . . . . . . . 4
1.5. The zeta point of view . . . . . . . . . . . . . . . . . . . . 5

Chapter 2. Examples 7
2.1. Examples where dim $X(\mathbf{C}) = 0$ . . . . . . . . . . . . . . 7
2.2. Examples where dim $X(\mathbf{C}) = 1$ . . . . . . . . . . . . . . 10
2.3. Examples where dim $X(\mathbf{C}) = 2$ . . . . . . . . . . . . . . 12

Chapter 3. The Chebotarev density theorem for a number field 15
3.1. The prime number theorem for a number field . . . . . . . 15
3.2. Chebotarev theorem . . . . . . . . . . . . . . . . . . . . . 17
3.3. Frobenian functions and frobenian sets . . . . . . . . . . . 20
3.4. Examples of $S$-frobenian functions and $S$-frobenian sets . . . . . . . . . . . . . . . . . . . . . 25

Chapter 4. Review of $\ell$-adic cohomology 31
4.1. The $\ell$-adic cohomology groups . . . . . . . . . . . . . . . 31
4.2. Artin's comparison theorem . . . . . . . . . . . . . . . . . 32
4.3. Finite fields : Grothendieck's theorem . . . . . . . . . . . 33
4.4. The case of a finite field : the geometric and the arithmetic Frobenius . . . . . . . . . . . . . . . . . . 34
4.5. The case of a finite field : Deligne's theorems . . . . . . . 35
4.6. Improved Deligne-Weil bounds . . . . . . . . . . . . . . . 36
4.7. Examples . . . . . . . . . . . . . . . . . . . . . . . . . . . 40
4.8. Variation with $p$ . . . . . . . . . . . . . . . . . . . . . . 42

CHAPTER 5. AUXILIARY RESULTS ON GROUP REPRESENTATIONS 45
5.1. Characters with few values . . . . . . . . . . . . . . . . . . 45
5.2. Density estimates . . . . . . . . . . . . . . . . . . . . . . 56
5.3. The unitary trick . . . . . . . . . . . . . . . . . . . . . . 59

CHAPTER 6. THE $\ell$-ADIC PROPERTIES OF $N_X(p)$ 65
6.1. $N_X(p)$ viewed as an $\ell$-adic character . . . . . . . . . . . . 65
6.2. Density properties . . . . . . . . . . . . . . . . . . . . . . 73
6.3. About $N_X(p) - N_Y(p)$ . . . . . . . . . . . . . . . . . . . 78

CHAPTER 7. THE ARCHIMEDEAN PROPERTIES OF $N_X(p)$ 83
7.1. The weight decomposition of the $\ell$-adic character $h_X$ . . . 83
7.2. The weight decomposition : examples and applications . . 90

CHAPTER 8. THE SATO-TATE CONJECTURE 101
8.1. Equidistribution statements . . . . . . . . . . . . . . . . . 101
8.2. The Sato-Tate correspondence . . . . . . . . . . . . . . . . 106
8.3. An $\ell$-adic construction of the Sato-Tate group . . . . . . . 111
8.4. Consequences of the Sato-Tate conjecture . . . . . . . . . . 114
8.5. Examples . . . . . . . . . . . . . . . . . . . . . . . . . . 121

CHAPTER 9. HIGHER DIMENSION: THE PRIME NUMBER THEOREM AND THE CHEBOTAREV DENSITY THEOREM 131
9.1. The prime number theorem . . . . . . . . . . . . . . . . . 131
9.2. Densities . . . . . . . . . . . . . . . . . . . . . . . . . . 134
9.3. The Chebotarev density theorem . . . . . . . . . . . . . . 136
9.4. Proof of the density theorem . . . . . . . . . . . . . . . . 138
9.5. Relative schemes . . . . . . . . . . . . . . . . . . . . . . 144

REFERENCES 147

INDEX OF NOTATIONS 157

INDEX OF TERMS 161

# Preface

The title of these lectures requires an explanation: what does $N_X(p)$ mean?

Answer: $N_X(p)$ is the number of solutions mod $p$ of a given family $X$ of polynomial equations in several unknowns, and with coefficients in $\mathbf{Z}$, so that it makes sense to reduce mod $p$, and count the solutions. For a fixed $X$, one wants to understand how $N_X(p)$ varies with $p$ : what is its size and its congruence properties? Can it be computed by closed formulae, by cohomology, and/or by efficient computer programs? What are the open problems?

These questions offer a good opportunity for reviewing several basic techniques in algebraic geometry, group representations, number theory, cohomology (both $\ell$-adic and standard) and modular forms.

This is why I chose this topic for two week-long courses at the National Center for Theoretical Sciences (NCTS) in Hsinchu, Taiwan, in July 2009 and April 2011. A group of people wrote up a set of notes based on my 2009 lectures, and I rewrote and expanded them. Here is the result of that rewriting–expanding.

There are nine chapters. The first four are preliminary, and short : they contain almost no proofs.

Chapter 1 gives an overview of the main theorems on $N_X(p)$ that will be discussed later, and Chapter 2 contains explicit examples, chosen for their simplicity and/or for aesthetic reasons.

Chapter 3 is about the Chebotarev density theorem, a theorem that is essential in almost everything done in Chapters 6 and 7; note in particular the "frobenian functions" and "frobenian sets" of §3.3 and §3.4.

Chapter 4 reviews the part of $\ell$-adic cohomology that will be used later.

Chapter 5 contains results on group representations that are difficult to find explicitly in the literature, for instance the technique consisting of computing Haar measures in a compact $\ell$-adic group by doing a similar computation in a real compact Lie group.

These results are applied in Chapter 6 in order to discuss the possible relations between two different families of equations $X$ and $Y$. Here is an example : suppose that $|N_X(p) - N_Y(p)| \leqslant 1$ for every large enough $p$; then there are only three possibilities :

i) $N_X(p) = N_Y(p)$ for every large enough $p$;

ii) there exists a non-zero integer $d$ such that $N_X(p) - N_Y(p) = (\frac{d}{p})$ for every large enough $p$;

iii) same as ii) with $(\frac{d}{p})$ replaced by $-(\frac{d}{p})$.

This looks mysterious at first, but if one transforms it into a statement on group characters, it becomes very simple.

Chapter 7 is about the archimedean properties of the $N_X(p)$ – a topic on which we know much less than in the $\ell$-adic case.

Chapter 8 is an introduction to the Sato-Tate conjecture, and its concrete aspects.

Chapter 9 gives an account of the prime number theorem, and of the Chebotarev density theorem, in higher dimension.

The text contains a few complementary results, usually written as exercises, with hints.

It is a pleasure to thank NCTS and its director Winnie Li for their hospitality and for their help during and after these lectures. I also thank K.S. Kedlaya and K. Ribet for their numerous corrections.

Paris, August 2011

Jean-Pierre Serre

# Conventions

The symbols $\mathbf{Z}, \mathbf{F}_p, \mathbf{F}_q, \mathbf{Q}, \mathbf{Q}_\ell, \mathbf{R}, \mathbf{C}, \mathbf{GL}_n, \mathbf{Sp}_{2n}$ have their usual meaning.

The cardinal number of a set $S$ is denoted by $|S|$.

$\mathbf{N} = \{0, 1, ...\}$ is the set of the cardinal numbers of the finite sets.

The symbols $\sqcup$ and $\bigsqcup$ denote disjoint unions.

If $X \subset Y$, the complement of $X$ in $Y$ is denoted by $Y - X$.

Positive means " $> 0$ or zero " (except in §4.6) ; it is almost always written as " $\geqslant 0$ " in order to avoid any confusion with "strictly positive".

If $A$ is a ring, $A^\times$ is the group of invertible elements of $A$.

The letters $\ell$ and $p$ are only used to denote primes (except for the $(p,q)$-terminology of Hodge types, which occurs in §8.2 and §8.3) ; most of the time, we assume $p \neq \ell$.

If $k$ is a field, $\overline{k}$ is an algebraic closure of $k$ and $k_s$ is the maximal separable extension of $k$ contained in $\overline{k}$ ; the Galois group $\mathrm{Gal}(k_s/k) = \mathrm{Aut}_k(\overline{k})$ is denoted by $\Gamma_k$.

If $X$ is a scheme over a commutative ring $A$, and if $A \to B$ is a homomorphism of commutative rings, $X(B)$ denotes the set of $B$-points of $X$, i.e. the set of $A$-morphisms of $\mathrm{Spec}\, B$ into $X$. The $B$-scheme deduced from $X$ by the base-change $\mathrm{Spec}\, B \to \mathrm{Spec}\, A$ is denoted by $X_{/B}$.

If $k$ is a field, a $k$-variety is a scheme of finite type over Spec $k$ ; it is not required that it is separated. However, the reader may make this reassuring assumption without losing much ; similarly, all schemes may be assumed to be quasi-projective and reduced.

A measure on a compact topological space $X$ is a Radon measure in the sense of [INT] (see also [Go 01]), i.e. a continuous linear form $\varphi \mapsto \mu(\varphi)$ on the Banach space of continuous functions on $X$. Most of the measures we consider are positive of mass 1 ; this means that $\mu(1) = 1$ and $\varphi \geqslant 0 \Rightarrow \mu(\varphi) \geqslant 0$.

The comparison symbols $\{O,\ o,\ \sim,\ \ll\}$ of analytic number theory have their usual meaning ; it is recalled the first time the symbol occurs.

# CHAPTER 1
# INTRODUCTION

## 1.1. Definition of $N_X(p)$ : the affine case

Let $f_\alpha(X_1, ..., X_n) \in \mathbf{Z}[X_1, ..., X_n]$ be a family of polynomials with integer coefficients. If $p$ is a prime number, let $N_f(p)$ be the number of solutions of the system of equations $f_\alpha(x_1, ..., x_n) = 0$ over the field $\mathbf{F}_p = \mathbf{Z}/p\mathbf{Z}$, i.e.

$$N_f(p) = |\{(x_1, ..., x_n) : x_i \in \mathbf{F}_p, f_\alpha(x_1, ..., x_n) = 0 \text{ in } \mathbf{Z}/p\mathbf{Z} \text{ for each } \alpha\}|.$$

This can be translated into the language of commutative algebra as follows. Let $(f_\alpha)$ be the ideal of $\mathbf{Z}[X_1, ..., X_n]$ generated by the polynomials $f_\alpha$. The quotient $A = \mathbf{Z}[X_1, ..., X_n]/(f_\alpha)$ is a ring of finite type over $\mathbf{Z}$. The points $x \in (\mathbf{F}_p)^n$ with $f_\alpha(x) = 0$ for all $\alpha$ correspond bijectively to the homomorphisms $A \to \mathbf{F}_p$, or equivalently to the maximal ideals[1] of $A$ with residue field $\mathbf{F}_p$; we are just counting such homomorphisms.

More generally, we shall be interested in the number of solutions of the equations $f_\alpha(x) = 0$ in a finite field of order $p^e$, for $e = 1, 2, ...$; this number will be denoted[2] by $N_f(p^e)$.

## 1.2. Definition of $N_X(p)$ : the scheme setting

Consider a scheme $X$ of finite type over $\mathbf{Z}$; this assumption means that $X$ has a finite covering by open subschemes of the form $\operatorname{Spec} A$ with $A$ of finite type over $\mathbf{Z}$. [In the affine case above, one has $X = \operatorname{Spec} A$, where $A = \mathbf{Z}[X]/(f_\alpha)$.]

We denote by $X_{/\mathbf{Q}}$ the generic fiber of $X \to \operatorname{Spec} \mathbf{Z}$, i.e. the $\mathbf{Q}$-algebraic variety deduced from $X$ by the base change $\mathbf{Z} \to \mathbf{Q}$.

The fiber $X_p$ of $X \to \operatorname{Spec} \mathbf{Z}$ at a prime $p \in \operatorname{Spec} \mathbf{Z}$ is a scheme over $\mathbf{F}_p$. We denote by $N_X(p)$ the number of its $\mathbf{F}_p$-points; equivalently :

$N_X(p)$ = number of closed points $x \in X$ whose residue field $\kappa(x)$ is such that $|\kappa(x)| = p$.

If $q = p^e$ is a power of $p$, with $e \geqslant 1$, we define similarly $N_X(q)$ as the cardinality of the set $X(\mathbf{F}_q)$ of the $\mathbf{F}_q$-points of $X$ (or of $X_p$ - that amounts

[1] Recall that, if $\mathfrak{m}$ is a maximal ideal of $A$, the quotient $A/\mathfrak{m}$ is a finite field, cf. e.g., [AC V-68, cor.1]. Equivalently, if $X$ is a scheme of finite type over $\mathbf{Z}$, a point $x$ of $X$ is closed if and only if its residue field $\kappa(x)$ is finite.

[2] Warning. One should not confuse $N_f(p^e)$ with the number $N_f(\operatorname{mod} p^e)$ of solutions of the equations $f_\alpha(x) = 0$ in the ring $\mathbf{Z}/p^e\mathbf{Z}$; for examples of computation of $N_f(\operatorname{mod} p^e)$, see the exercises of §1.3, §2.1.2 and §2.2.3.

to the same). Recall that an element of $X(\mathbf{F}_q)$ is a morphism of $\operatorname{Spec} \mathbf{F}_q$ into $X$ ; one may view it as a pair $(x, \varphi)$ where $x$ is a closed point of $X$ and $\varphi$ is an embedding of $\kappa(x)$ into $\mathbf{F}_q$.

As explained in the Preface, the goal of the present lectures is to describe how $N_X(p)$ varies with $p$, and in particular to relate this variation to the topology of the analytic space $X(\mathbf{C})$ made up of the complex points of $X$. [Note that, in the setting of §1.1, $X(\mathbf{C})$ is the set of all $x \in \mathbf{C}^n$ such that $f_\alpha(x) = 0$ for every $\alpha$.]

*Remarks about $N_X(p^e)$.*

1) $N_X(p^e)$ is additive in $X$ : if $X$ is a disjoint union of subschemes $X_i$, we have $N_X(p^e) = \sum N_{X_i}(p^e)$ for every prime $p$ and every $e \geqslant 1$.

2) $N_X(p^e)$ depends only on the reduced scheme $X^{\mathrm{red}}$ associated with $X$ : nilpotent elements play no role.

3) If two schemes $X$ and $Y$ become isomorphic over $\mathbf{Q}$, i.e. if $X_{/\mathbf{Q}} \simeq Y_{/\mathbf{Q}}$, then there exists a prime $p_0$ such that $N_X(p^e) = N_Y(p^e)$ for every $p \geqslant p_0$ and every $e \geqslant 1$. This shows that the knowledge of the $\mathbf{Q}$-variety $X_{/\mathbf{Q}}$ is enough to determine the $N_X(p^e)$, for all $p$ but finitely many.

4) There is no need to assume $e \geqslant 1$ : there is a reasonable definition of $N_X(p^e)$ for every $e \in \mathbf{Z}$, see the end of §1.5; note however that, when $e \leqslant 0$, $N_X(p^e)$ does not usually belong to $\mathbf{N}$, nor even to $\mathbf{Z}$, but it belongs to $\mathbf{Z}[1/p]$.

Remarks 1), 2) and 3) make possible several dévissage arguments; according to our needs, we may for instance assume that $X$ is affine, or separated, or projective, or smooth (either over $\mathbf{Z}$ or over $\mathbf{Q}$), etc.

*Exercise.* Let $X$ be a scheme of finite type over $\mathbf{Z}$. Show that there exists a finite set of polynomials $f = (f_\alpha)$, as in §1.1, such that $N_X(p^e) = N_f(p^e)$ for every prime $p$ and every integer $e$.
[Hint. Use the fact that $X$ is noetherian to show that there exists a decreasing sequence of closed subschemes $X = X_0 \supset X_1 \supset ... \supset X_n = \varnothing$ such that each $X_i - X_{i+1}$ is affine of finite type over $\mathbf{Z}$. If $Y$ is the disjoint union of the $X_i - X_{i+1}$, then $Y$ is affine, and $N_X(p^e) = N_Y(p^e)$ for every $p, e$.]

## 1.3. How large is $N_X(p)$ when $p \to \infty$?

Here are some of the results that we shall discuss later (mainly in Chapter 7). The first one is very simple; it tells us that the empty set can be detected by counting its points mod $p$ :

**Theorem 1.1.** $X(\mathbf{C}) = \varnothing \iff N_X(p) = 0$ *for every large enough* $p$.

*Remark.* Note first that $X(\mathbf{C}) = \varnothing$ is equivalent to $X_{/\mathbf{Q}} = \varnothing$. Suppose that this does not happen, and let $P_X$ be the set of $p$ such that $N_X(p) \neq 0$. It is known (see [Ax 67], as well as §7.2.4) that $P_X$ has a density[3], which is a strictly positive rational number. The same is true for the set of $p$ with $N_X(p^2) \neq 0$; same for $N_X(p^3) \neq 0$, etc.

The next theorem (proved in §7.2.1) relates the asymptotic behavior of $N_X(p)$ with the complex dimension of $X(\mathbf{C})$ (which is the same as dim $X_{/\mathbf{Q}}$) :

**Theorem 1.2.** *Let* $d \geqslant 0$.

(a) $\dim X(\mathbf{C}) \leqslant d \iff N_X(p) = O(p^d)$ *when* $p \to \infty$.

(b) *Assume* $\dim X(\mathbf{C}) \leqslant d$. *Let* $r$ *be the number of* $\mathbf{C}$*-irreducible components of dimension* $d$ *of* $X(\mathbf{C})$. *Then*

$$\limsup_{p\to\infty} \frac{N_X(p)}{p^d} = r.$$

(c) *Assume* $\dim X(\mathbf{C}) \leqslant d$. *Let* $r_0$ *be the number of* $\mathbf{Q}$*-irreducible components of dimension* $d$ *of* $X_{/\mathbf{Q}}$. *Then*

$$\sum_{p \leqslant x} N_X(p) = r_0 \frac{x^{d+1}}{\log(x^{d+1})} + O(x^{d+1}/(\log x)^2) \text{ when } x \to \infty.$$

Recall that the $O$-notation in (a) means that there exist a prime $p_0$ and a number $C > 0$ such that

$$N_X(p) \leqslant Cp^d \quad \text{for all } p \geqslant p_0.$$

*Remark.* Theorem 1.2 shows that the asymptotic properties of the function $p \mapsto N_X(p)$ detect the dimension $d$ of $X(\mathbf{C})$, the number of its irreducible components of dimension $d$, and the number of $\mathbf{Q}$-irreducible components of $X_{/\mathbf{Q}}$ of dimension $d$.

*Example.* The equation $x^2 + y^2 = 0$ represents the union of two lines in the affine plane, with slopes $i$ and $-i$. We have $d = 1, r = 2, r_0 = 1$ and

$$N_X(p^e) = \begin{cases} p^e & \text{if} \quad p = 2 \\ 2p^e - 1 & \text{if} \quad p^e \equiv 1 \pmod 4 \\ 1 & \text{if} \quad p^e \equiv 3 \pmod 4). \end{cases}$$

[3] a "natural density", in the sense defined later (§3.1.3); we shall not be interested in the weaker notion of "Dirichlet density", which is mostly useful in the equal-characteristic case.

This example shows that " lim sup " cannot be replaced by " lim " in Theorem 1.2.(b).

*Remark.* In §7.2.1 we shall prove Theorem 1.2 in a refined form, and with a better error term (the same as the one occurring in the Chebotarev density theorem, cf. §3.2.3).

*Exercise.* Let $N(\bmod p^e)$ be the number of solutions of $x^2 + y^2 = 0$ in the ring $\mathbf{Z}/p^e\mathbf{Z}$. Show that :

i) $N(\bmod 2^e) = 2^e$.

ii) If $p \equiv 1 \pmod 4$, then $N(\bmod p^e) = (e+1)p^e - ep^{e-1}$.

iii) If $p \equiv 3 \pmod 4$, then $N(\bmod p^e) = p^e$ if $e$ is even and $N(\bmod p^e) = p^{e-1}$ if $e$ is odd.

## 1.4. More properties of $p \mapsto N_X(p)$

The next theorem is a kind of *rigidity property* of the function $p \mapsto N_X(p)$. It will be proved in §6.1.3. Here again, the proof uses the Chebotarev density theorem. But it also depends in an essential way on the properties of $\ell$-adic cohomology, due to Grothendieck; we shall recall them in Chapter 4.

**Theorem 1.3.** *Let $X, Y$ be two schemes of finite type over $\mathbf{Z}$. Assume that $N_X(p) = N_Y(p)$ for a set of primes $p$ of density 1. Then there exists a prime number $p_0$ such that $N_X(p^e) = N_Y(p^e)$ for all $p \geqslant p_0$ and all $e \geqslant 1$.*

[If $X$ and $Y$ are separated, the "density 1" hypothesis can be replaced by " density $> 1 - 1/B^2$ ", where $B$ depends only on the cohomology of the spaces $X(\mathbf{C})$ and $Y(\mathbf{C})$, cf. Theorem 6.17.]

The same method also gives the following curious-looking result (cf. §6.1.2) :

**Theorem 1.4.** *Let $X$ be a scheme of finite type over $\mathbf{Z}$. Let $a$ and $m$ be integers with $m \geqslant 1$. The set of primes $p$ such that $N_X(p) \equiv a \pmod m$ has a density, which is a rational number. If $X$ is separated*[4] *and $a$ is equal to the Euler-Poincaré characteristic $\chi(X(\mathbf{C}))$ of $X(\mathbf{C})$, that density is $> 0$.*

**Corollary 1.5.** *For every $m \geqslant 1$, the set of $p$ such that*

$$N_X(p) \equiv \chi(X(\mathbf{C})) \pmod m$$

*is infinite.*

Let us recall what the Euler-Poincaré characteristic is. If $T$ is a locally compact space, let $H^i(T, \mathbf{Q})$ be the $i$-th cohomology group of $T$ with coefficients in $\mathbf{Q}$; for $i = 0$, the $\mathbf{Q}$-vector space $H^i(T, \mathbf{Q})$ is the space of locally

[4]This assumption insures that $X(\mathbf{C})$ is locally compact, so that its cohomology with compact support is well defined.

constant functions $T \to \mathbf{Q}$. We also have $H^i_c(T, \mathbf{Q})$, the cohomology *with compact support*; here $H^0_c(T, \mathbf{Q})$ is the $\mathbf{Q}$-vector space of locally constant functions $T \to \mathbf{Q}$ that vanish outside a compact subspace of $T$.

If the $H^i$ and the $H^i_c$ are finite-dimensional vector spaces and vanish for $i$ large, we may define the Euler-Poincaré characteristic of $T$ by the usual formula :

$$\chi(T) = \sum_{i \geqslant 0} (-1)^i \dim H^i(T, \mathbf{Q}),$$

together with its variant with compact support :

$$\chi_c(T) = \sum_{i \geqslant 0} (-1)^i \dim H^i_c(T, \mathbf{Q}).$$

These definitions apply[5] to the locally compact space $T = X(\mathbf{C})$. Moreover, by a theorem of Laumon (cf. [La 81]), we have $\chi(T) = \chi_c(T)$. Hence, in Theorem 1.4, we may use at will[6] either $\chi$ or $\chi_c$.

*Remark.* The formula $\chi(T) = \chi_c(T)$ would become false if $T = X(\mathbf{C})$ were replaced by $T = X(\mathbf{R})$. For instance, if $X$ is the affine line and $T = X(\mathbf{R}) = \mathbf{R}$, we have $\chi(T) = 1$ and $\chi_c(T) = -1$. More generally, Poincaré duality shows that, for a real orientable manifold $V$ of (real) dimension $d$, we have $\chi(V) = (-1)^d \chi_c(V)$.

## 1.5. The zeta point of view

It is often convenient to pack the information given by $N_X(p^e)$ into one single object : the *zeta function* $\zeta_X(s)$ of the scheme $X$, i.e. the Dirichlet series $\sum a_n n^{-s}$ defined by the infinite product

$$\zeta_X(s) = \prod_{x \in \underline{X}} \frac{1}{1 - |x|^{-s}},$$

where $x$ runs through the set $\underline{X}$ of closed points of $X$ and $|x|$ is the number of elements of the residue field $\kappa(x)$. The product converges absolutely for $\mathrm{Re}(s) > \dim X$, see e.g. [Se 65] ; here $\dim X$ is the dimension of the scheme $X$, not that of $X(\mathbf{C})$; for instance $\dim \mathrm{Spec}\, \mathbf{Z} = 1$.

---

[5] Indeed, it is enough to prove this when $X$ is quasi-projective (affine would be enough), in which case the triangulation theorem of analytic spaces shows that $T$ is homeomorphic to $K - L$, where $K$ is a finite simplicial complex and $L$ is a closed subcomplex of $K$. (A proof, in the more general setting of "semi-algebraic sets", can be found in [BCR 98, §9.2].) This implies that the abelian groups $H^i(T, \mathbf{Z})$ and $H^i_c(T, \mathbf{Z})$ are finitely generated and are 0 for $i > \dim_{\mathrm{top}} T = 2 \dim X(\mathbf{C})$; the same is true for the $\mathbf{Q}$-vector spaces $H^i(T, \mathbf{Q})$ and $H^i_c(T, \mathbf{Q})$.

[6] However, it is $\chi_c$, rather than $\chi$, that occurs naturally in the proofs, cf. Chap.6.

A standard computation shows that the Dirichlet series $\zeta_X(s)$ can be written as an Euler product

$$\zeta_X(s) = \prod_p \zeta_{X,p}(s),$$

where

$$\zeta_{X,p}(s) = \exp(\sum_{e=1}^{\infty} N_X(p^e)p^{-es}/e).$$

Hence $\zeta_X(s)$ is determined by the $N_X(p^e)$. Conversely, $N_X(p^e)$ can be recovered from the coefficients $a_n$ of $\zeta_X(s)$, by expanding the identity

$$N_X(p)t + N_X(p^2)t^2/2 + \cdots = \log(1 + a_p t + a_{p^2}t^2 + \cdots),$$

where $t$ is an indeterminate. For instance :

$$N_X(p) = a_p,\ N_X(p^2) = 2a_{p^2} - a_p^2,\ N_X(p^3) = 3a_{p^3} - 3a_p a_{p^2} + a_p^3.$$

In what follows, we will mostly work with $N_X(p^e)$, but once in a while we shall also mention the zeta aspect.

*Definition of $N_X(p^e)$ for $e \leqslant 0$.*

By *Dwork's rationality theorem* ([Dw 58]), proved by $p$-adic methods (and reproved $\ell$-adically by Grothendieck, see §4.4), *$\zeta_{X,p}(s)$ is a rational function of $p^{-s}$*. More precisely, there is a product decomposition

$$\zeta_{X,p}(s) = \prod_{z \in \mathbf{C}^\times} (1 - zp^{-s})^{n(z)},$$

where the integers $n(z)$ are 0 for every $z \in \mathbf{C}^\times$ except a finite number. This is equivalent to saying that

$$(*) \qquad N_X(p^e) = \sum_z n(z)z^e \quad \text{for every } e \geqslant 1.$$

By a theorem of Deligne, the $z$ that occur with a non-zero coefficient $n(z)$ are " $p$-Weil integers ", see §4.5 ; moreover, one has $n(z) = n(z')$ if the algebraic numbers $z$ and $z'$ are conjugates over $\mathbf{Q}$.

If $e \in \mathbf{Z}$ is $\leqslant 0$, one may take (*) as the *definition* of $N_X(p^e)$ ; hence it makes sense to write $N_X(p^{-1})$, or $N_X(p^0)$, and the reader can check that these numbers enjoy most of the properties of the standard $N_X(p^e)$ for $e \geqslant 1$ ; the main differences are :

- $N_X(p^0)$ belongs to $\mathbf{Z}$, but not always to $\mathbf{N}$ (it is equal to the Euler characteristic of $\overline{X}_p$, cf. §4.3) ;
- for $e < 0$, $N_X(p^e)$ belongs to $\mathbf{Z}[\frac{1}{p}]$, but not always to $\mathbf{Z}$.

# Chapter 2
# Examples

We collect here a few examples where dim $X(\mathbf{C})$ is equal to 0, 1 or 2. Most of them will be reconsidered later from the viewpoint of étale cohomology, cf. §4.6.

## 2.1. Examples where dim $X(\mathbf{C}) = 0$

We take $f \in \mathbf{Z}[x]$ with $f \neq 0$ and consider $X = \operatorname{Spec} \mathbf{Z}[x]/(f)$, so that $N_X(p)$ is the number of solutions in $\mathbf{Z}/p\mathbf{Z}$ of $f(x) \equiv 0 \pmod p$.

### 2.1.1. A quadratic equation

Take $f = x^2 + 1$. We have

$$N_X(p) = \begin{cases} 1 & \text{if } p = 2\ ; \\ 2 & \text{if } p \equiv 1 \pmod 4\ ; \\ 0 & \text{if } p \equiv -1 \pmod 4\ . \end{cases}$$

By Dirichlet's theorem on arithmetic progressions, each of the two cases $p \equiv 1 \pmod 4$ and $p \equiv -1 \pmod 4$ occurs with density $\frac{1}{2}$. However, Sarnak and Rubinstein have shown (assuming the truth of some rather strong conjectures[1]) that there are quite often more primes $\equiv -1 \pmod 4$ than primes $\equiv 1 \pmod 4$. (For the precise definition of "quite often", see their paper [RS 94].)

### 2.1.2. A typical cubic equation

Take $f = x^3 - x - 1$. This polynomial has discriminant $4 - 27 = -23$. When $p = 23$, the equation $f = 0$ has a double root mod $p$ and $N_X(23)$ is equal to 2. For $p \neq 23$, we have (see [Bl 52] and [Se 03]) :

$$N_X(p) = \begin{cases} 1 & \text{if } \left(\frac{p}{23}\right) = -1; \\ 3 & \text{if } \left(\frac{p}{23}\right) = 1 \ \&\ p \text{ is represented by the binary form} \\ & \qquad a^2 + ab + 6b^2; \\ 0 & \text{if } \left(\frac{p}{23}\right) = 1 \ \&\ p \text{ is represented by the binary form} \\ & \qquad 2a^2 + ab + 3b^2. \end{cases}$$

[1] They assume that the non-trivial zeros of the Dirichlet functions are on the line $\operatorname{Re}(s) = \frac{1}{2}$, are simple, and that their imaginary parts (normalized to be $> 0$) are $\mathbf{Q}$-linearly independent.

These three sets of primes have density 1/2, 1/6 and 1/3 respectively.

*Remark.* The result can also be expressed in a more compact form by introducing the power series $F_{23} = \sum_{n=1}^{\infty} a_n q^n$ defined by the formula :

$$F_{23} = \frac{1}{2}\Big(\sum_{a,b\in\mathbf{Z}} q^{a^2+ab+6b^2} - \sum_{a,b\in\mathbf{Z}} q^{2a^2+ab+3b^2}\Big) = q\prod_{n=1}^{\infty}(1-q^n)(1-q^{23n})$$
$$= q - q^2 - q^3 + q^6 + q^8 - q^{13} - q^{16} + q^{23} + \cdots,$$

which is a cusp form of level 23 and weight 1, cf. e.g. [Fr 28, p.472].

It can be proved that

$$N_X(p) = a_p + 1$$

for every $p$. Moreover, the zeta function $\zeta_X(s)$ of $X$ is equal to $L_{23}(s)\zeta(s)$, where $\zeta(s)$ is the standard zeta function (i.e. that of Spec $\mathbf{Z}$), and $L_{23}(s)$ is the Dirichlet series

$$L_{23}(s) = \sum_{n\geqslant 1} a_n n^{-s} = \prod_p \frac{1}{1 - a_p p^{-s} + \left(\frac{p}{23}\right) p^{-2s}}$$

whose coefficients $a_n$ are the same as those of the power series $F_{23}$. See [Se 02, §5.3] for an interpretation in terms of Artin L-functions.

*Exercises.*

1) Let $N_X(\text{mod}\, p^e)$ be the number of solutions of $x^3 - x - 1 = 0$ in $\mathbf{Z}/p^e\mathbf{Z}$. Show that :

i) $N_X(\text{mod}\, p^e) = N_X(p)$ if $p \neq 23$.

ii) $N_X(\text{mod}\, p^e) = 1$ if $p = 23$ and $e > 1$.

2) Let $N_X(p^e)$ be the number of solutions of $x^3 - x - 1 = 0$ in a field with $p^e$ elements. Show that $N_X(p^e) = 3$ if $N_X(p) = 1$ and $e$ is even, or $N_X(p) = 0$ and $e$ is divisible by 3. Show that $N_X(p^e) = N_X(p)$ otherwise.

### 2.1.3. Another cubic equation

Take $f = x^3 + x + 1$. This polynomial has discriminant $-4-27 = -31$. The results are almost the same[2] as those for $x^3 - x - 1$, the binary quadratic forms $a^2+ab+6b^2$ and $2a^2+ab+3b^2$ with discriminant $-23$ being replaced by $a^2 + ab + 8b^2$ and $2a^2 + ab + 4b^2$, which have discriminant $-31$. Here also we have $N_X(p) = a_p + 1$, where $a_p$ is the $p$-th coefficient of the cusp form

$$F_{31} = \frac{1}{2}\Big(\sum_{a,b\in\mathbf{Z}} q^{a^2+ab+8b^2} - \sum_{a,b\in\mathbf{Z}} q^{2a^2+ab+4b^2}\Big).$$

[2] The main point is that $h(-23) = h(-31) = 3$, i.e. the quadratic fields of discriminant $-23$ and $-31$ have class number 3, cf. §3.3.3.3.

The main difference is that it is not possible to write $F_{31}$ as a product $q\prod_{n\geqslant 1}(1-q^n)^{c_n}$ with bounded exponents $c_n$ (for a more precise statement, giving the value of $\limsup \frac{1}{n}\log|c_n|$, see exercise below).

*Exercise.*

a) Show that $\begin{pmatrix} -6 & 1 \\ -31 & 5 \end{pmatrix}$ is an element of order 3 of the modular group $\Gamma_0(31)$, that fixes the point $z_0 = \frac{11+i\sqrt{3}}{2\cdot 31}$ of the upper-half plane $\mathrm{Im}(z) > 0$ .

b) Show that the modular form $z \mapsto F_{31}(e^{2\pi i z})$ vanishes on the $\Gamma_0(31)$-orbit of $z_0$, and at no other point of the upper-half plane; show that it is non-zero when $\mathrm{Im}(z) > \mathrm{Im}(z_0)$.

c) Use b) to prove that the radius of convergence of the $q$-series $\log(\frac{1}{q}F_{31}(q))$ is equal to $e^{-\pi\sqrt{3}/31}$.

d) Show that, if one writes $F_{31}$ as $q\prod(1-q^n)^{c_n}$ with $c_n \in \mathbf{Z}$ (which is possible in a unique way), then $\limsup \frac{1}{n}\log|c_n| = \pi\sqrt{3}/31$.

### 2.1.4. Computational problems

If $f$ is given, the problem of computing the corresponding $N_X(p)$ for large $p$ is a $P$-problem : there is a deterministic algorithm that solves it in time $O((\log p)^A)$ for a suitable exponent $A$. The method is simple[3] : it consists in computing the gcd $g(x)$ of $f(x)$ and $x^p - x$ in $\mathbf{F}_p[x]$ ; the degree of $g(x)$ is $N_X(p)$. This gives an exponent $A$ equal to 3 ; using "fast multiplication" (cf. [Kn 81, §4.3.3]) one can bring $A$ down to $2+\varepsilon$ for every $\varepsilon > 0$.

There is a special case where one can do better. When the Galois group of $f$ is abelian, the roots of $f$ belong to a cyclotomic field $\mathbf{Q}(z_m)$, where $z_m$ is a primitive $m$-th root of unity. In that case, the value of $N_X(p)$ depends only[4] on the value of $p$ mod $m$ ; since this value can be computed in time $O(\log p)$, we can take $A = 1$.

*Problem. Is this the only case where the exponent $A$ can be taken $< 2$?*

For instance, in the two cubic cases given above, can one prove that there does not exist any deterministic algorithm computing $N_X(p)$ in time $O((\log p)^A)$ with $A < 2$?

Another natural question is :

*Problem. Compute the roots of $f$ mod $p$ in polynomial time.*

This is easily done if one accepts probabilistic algorithms (see e.g. [Kn 81, §4.6.2]), but no deterministic polynomial time algorithm seems

---

[3] As J-F. Mestre pointed out to me, this method was already known to Libri and to Galois around 1830, cf. [Ga 30].

[4] I am assuming here that $p$ does not divide the discriminant of $f$.

to be known in the general case, except when $\deg f = 2$, thanks to a theorem of Schoof [Sc 85, §4], and in a few other cases (roots of unity of prime order, cf. [Pi 90]). I do not know what the situation is for the two cubic equations written above.

## 2.2. Examples where $\dim X(\mathbf{C}) = 1$

### 2.2.1. Genus 0

Take for $X$ the conic in the projective plane[5] $\mathbf{P}_2$ defined by the homogeneous equation

$$x^2 + y^2 + z^2 = 0.$$

We have $N_X(p) = p + 1$ for every prime $p$ : this is obvious for $p = 2$ since the equation is equivalent to $x + y + z = 0$, hence has the three solutions (1,0,0), (0,1,0) and (0,0,1). For $p > 2$, the conic $X_p$ is smooth, and has an $\mathbf{F}_p$-rational point (by Chevalley-Warning, or by a direct argument), hence is isomorphic to $\mathbf{P}_1$, hence has $p + 1$ rational points.
[Alternative proof : use Weil's bound $|N_X(p) - (p+1)| \leqslant 2gp^{\frac{1}{2}}$, with $g = 0$.]
The zeta function of $X$ is $\zeta_X(s) = \zeta(s)\zeta(s-1)$.

### 2.2.2. Genus 1 with complex multiplication

Consider the elliptic curve $X$ in $\mathbf{P}_2$ given by the affine equation

$$y^2 = x^3 - x.$$

This curve has good reduction outside 2; its conductor is $2^5$, cf. [Cr 97, p.111, case 32A2(A)]. It has complex multiplication : its $\overline{\mathbf{Q}}$-endomorphism ring is the ring $\mathbf{Z}[i]$ of Gaussian integers, with $i$ acting as $(x, y) \mapsto (-x, iy)$. One finds that $N_X(p)$, for $p > 2$, is given by

$$N_X(p) = p + 1 - a_p,$$

where $a_p$ is as follows :

• if $p \equiv -1 \pmod 4$, we have $a_p = 0$, so that $N_X(p)$ is $p + 1$, as if the curve had genus 0 (but $N_X(p^2)$ is not $p^2 + 1$ : it is $p^2 + 2p + 1$);

• if $p \equiv 1 \pmod 4$, we can write $p$ as $\pi\overline{\pi}$ with $\pi \in \mathbf{Z}[i]$; hence $p = u^2 + v^2$ if $\pi = u + vi$ with $u, v \in \mathbf{Z}$. We can choose $\pi$ in a unique way[6] (up to conjugation) such that $\pi \equiv 1 \pmod{(1+i)^3}$. Then $a_p = 2u = \pi + \overline{\pi}$. Moreover, $N_X(p^e)$ is equal to $p^e + 1 - (\pi^e + \overline{\pi}^e)$ for every $e \geqslant 1$.

[5] Here - and elsewhere too - we use Bourbaki's notation : the $n$-dimensional projective space is denoted by $\mathbf{P}_n$, and not by $\mathbf{P}^n$ which would suggest that it is the $n$-th-power of $\mathbf{P}^1$.

[6] Every fractional ideal $\mathfrak{a}$ of $\mathbf{Z}[i]$ prime to $\lambda = 1 + i$ has a unique generator $\pi_{\mathfrak{a}}$ such that $\pi_{\mathfrak{a}} \equiv 1 \pmod{\lambda^3}$. The map $\mathfrak{a} \mapsto \pi_{\mathfrak{a}}$ is the Hecke character associated with the

### 2.2.3. Genus 1 without complex multiplication

Consider the elliptic curve $X$ in $\mathbf{P}_2$ given by the affine equation

$$y^2 - y = x^3 - x^2.$$

This curve has good reduction outside $p = 11$; its conductor is 11, cf. [Cr 97, p.110, case 11A3(A)].

To compute $N_X(p)$ we use the modular form

$$F_{11}(q) = q \prod_{n=1}^{\infty} (1 - q^n)^2 (1 - q^{11n})^2 = \sum_{n=1}^{\infty} a_n q^n,$$

which is a cusp form of level 11 and weight 2, see e.g. [Fr 28, p.432]. Here again, one can prove (Eichler, Shimura, cf. [Sh 66]) the formula

$$N_X(p) = p + 1 - a_p.$$

*Example.* The $q$-expansion of $F_{11}$ is $q - 2q^2 - q^3 + 2q^4 + q^5 + \cdots$; hence $a_2 = -2, a_3 = -1, a_5 = 1$ and $N_X(p) = 5$ for $p = 2, 3, 5$.[7]

*Zeta function.* We have

$$\zeta_X(s) = \frac{\zeta(s)\zeta(s-1)}{L(s)},$$

with

$$L(s) = \sum a_n \cdot n^{-s} = \prod_p \frac{1}{1 - a_p p^{-s} + \varepsilon(p)\, p^{1-2s}},$$

where the $a_n$'s are the same as above and $\varepsilon(p)$ is equal to 0 if $p = 11$ and to 1 if $p \neq 11$.

*Remark.* This kind of relation between zeta functions and modular forms is a special case of the "modularity conjecture", started by Taniyama in 1955,

---

elliptic curve $X$, cf. [De 53]. [Hint. The kernel of the action of $\lambda^3$ on $X$ is made up of the following eight points : the 2-division points ($x = 0, 1, -1, \infty$) and the four points with $x = i, -i$. Hence, if $p$ splits in $\mathbf{Q}(i)$, the corresponding Frobenius endomorphism, viewed as an element $\pi$ of $\mathbf{Z}[i]$, fixes that kernel, and we have $\pi \equiv 1 \pmod{\lambda^3}$.]

[7] Note that $N_X(p)$ is divisible by 5 for every $p \neq 11$, since the group $X(\mathbf{F}_p)$ contains a subgroup of order 5, namely the one with $(x, y) = (0,0), (0,1), (1,0), (1,1), (\infty, \infty)$. When $p = 11$, it is $N_X(p) - 1$ that is divisible by 5 : one has to remove the double point of the cubic in order to get an algebraic group (viz. the multiplicative group $\mathbf{G}_m$).

There is also an explicit formula for the value mod 5 of $N_X(p)/5$ when $p \neq 11$, namely $N_X(p)/5 \equiv (p-1)\alpha(p) \pmod 5$, where $\alpha$ is the unique homomorphism of $(\mathbf{Z}/11\mathbf{Z})^\times$ into $\mathbf{Z}/5\mathbf{Z}$ such that $\alpha(2) = 1$. [Hint. Use the fact that there exists a curve isogenous to $X$ that contains $\mu_5 \times \mathbf{Z}/5\mathbf{Z}$, namely the curve $X_0(11)$. For a different proof, and a generalization, see [Maz 78, p.139].]

made more precise by Weil in 1966[8] and eventually proved by Wiles and others, cf. [Wi 95] and [BCDT 01].

*Exercise.* Let $N_X(\bmod p^e)$ be the number of points of the projective curve $X$ in the ring $\mathbf{Z}/p^e\mathbf{Z}$. Show that :

i) $N_X(\bmod p^e) = p^{e-1}N_X(p)$ if $p \neq 11$.

ii) $N_X(\bmod p^e) = p^e - p^{e-1}$ if $p = 11$ and $e > 1$.

### 2.2.4. Computation of $N_X(p)$ for large $p$ when $\dim X(\mathbf{C}) = 1$

It is known that, for a fixed curve $X$ over $\mathbf{Q}$, the computation of $N_X(p)$ can be done in polynomial time with respect to $\log p$. This has been proved by Schoof [Sc 85] when the genus of $X$ is 1, and by Pila [Pi 90] in the general case; see also [KS08] for more practical aspects of the computation of $N_X(p)$ in the case of hyperelliptic curves.The case of varieties of higher dimension is open, cf. [CL 07] and [CE 11, Epilogue].

## 2.3. Examples where $\dim X(\mathbf{C}) = 2$

### 2.3.1. Affine quadratic cone

In affine 3-space, consider the quadratic cone $X$ defined by the equation $x^2 = yz$. One has $N_X(p^e) = p^{2e}$ for every prime $p$ and every $e \geqslant 1$, as if $X$ were isomorphic to affine 2-space (which it is not).

### 2.3.2. Quadrics in 3-space

Take for $X$ the quadric in $\mathbf{P}_3$ defined by the homogeneous equation

$$ax^2 + by^2 + cz^2 + dt^2 = 0$$

where $a, b, c, d$ are non-zero integers. Over $\mathbf{C}$, such a surface is isomorphic to $\mathbf{P}_1 \times \mathbf{P}_1$. Over a finite field $\mathbf{F}_q$ of characteristic $\neq 2$, this is true if and only if $abcd$ is a square (assuming that $abcd \neq 0$ in $\mathbf{F}_q$). In that case the number of the $\mathbf{F}_q$-points is $q^2 + 2q + 1$.

If $abcd$ is not a square in $\mathbf{F}_q$, the quadric is isomorphic to the "Weil's restriction of scalars" of the projective line $\mathbf{P}_1$ for the quadratic field extension $\mathbf{F}_{q^2}/\mathbf{F}_q$ ; the number of its $\mathbf{F}_q$-points is $q^2 + 1$.

---

[8] The years 1965-1967 were an especially favorable period for Number Theory : besides Weil's paper [We 67], and the Sato-Tate conjecture ([Ta 65]), there was the launching of Langlands program [La 67] and the introduction of motives by Grothendieck ([CS 03, pp.173-175]). It was already suspected at that time that these daring theories are but the different facets of the same mathematical object. Half a century later, a lot of progress has been made by Deligne, Faltings, Wiles, Taylor and others, but we still do not know exactly how the pieces fit together.

### 2.3.3. Rational surfaces

More generally, consider a smooth projective geometrically[9] irreducible surface $V$ over a finite field $k$ with $|k| = q$; assume that $V$ is *geometrically rational*, i.e. that it becomes birationally isomorphic to $\mathbf{P}_2$ after the ground field extension $\overline{k}/k$. Let *NS* be the Néron-Severi group of $V_{/\overline{k}}$. The Galois group of $\overline{k}/k$ acts on *NS*, and we have (cf. [Man 86, §27]) :

$$N_V(q) = q^2 + q\,\mathrm{Tr}(\sigma_q) + 1,$$

where $\mathrm{Tr}(\sigma_q)$ is the trace of the Frobenius element $\sigma_q \in \mathrm{Gal}(\overline{k}/k)$ acting on the free $\mathbf{Z}$-module *NS*.

In the case where $V$ is a smooth quadric, we have $NS = \mathbf{Z}^2$, $\sigma_q$ acts either trivially or by permuting the two factors, so that its trace is either 2 or 0, and we recover example 2.3.2.

The case where $V$ is a smooth cubic surface in $\mathbf{P}_3$ (due to Weil [We 54, p.588] - see also [Man 86, §27]) is particularly interesting. We then have $\mathrm{rank}(NS) = 7$. The 27 lines of the cubic surface give 27 elements of *NS*. The group of automorphisms of the incidence graph of these lines is isomorphic to $\mathrm{Weyl}(E_6)$ = Weyl group of the root system $E_6$. The action of $\mathrm{Gal}(\overline{k}/k)$ on these lines gives a homomorphism $\mathrm{Gal}(\overline{k}/k) \to \mathrm{Weyl}(E_6)$ that is well defined up to conjugation. In particular, we have a Frobenius conjugacy class $\sigma_q$ in $\mathrm{Weyl}(E_6)$, and the formula for $N_V(q)$ can be written as

$$N_V(q) = q^2 + (1 + a)q + 1,$$

where $a$ is the trace of $\sigma_q$ acting by the reflection representation of $\mathrm{Weyl}(E_6)$, which is of dimension 6.

Note that $a$ can only take the values -3, -2, -1, 0, 1, 2, 3, 4, 6, as one sees by looking at the character table of $\mathrm{Weyl}(E_6)$, cf. [ATLAS, p.27]. Hence :

$$q^2 - 2q + 1 \leqslant N_V(q) \leqslant q^2 + 7q + 1.$$

*Remark.* For a given $q$ one may ask what the possible values of $a$ in $\{-3, -2, -1, 0, 1, 2, 3, 4, 6\}$ are. This does not seem to be known. However Swinnerton-Dyer ([Sw 10]) has shown that the minimal value $a = -3$ is always possible, and that the maximal value $a = 6$ is possible provided $q \neq 2, 3, 5$.

---

[9] It is customary to say that a $k$-scheme $X$ has *geometrically* a property $P$ if the $\overline{k}$-scheme $X_{/\overline{k}}$ has property $P$ (as if geometry could only be done over algebraically closed fields).

*Exercise.* Let $V$ be a smooth cubic surface in $\mathbf{P}_3$ over $\mathbf{F}_q$; define $a \in \mathbf{Z}$ by the formula $N_V(q) = q^2 + (1+a)q + 1$, as above.

a) If $q = 2$, show that $a = 6$ is impossible because $V$ would have strictly more $\mathbf{F}_q$-points than $\mathbf{P}_3$.

b) If $q = 4$, show that the smooth cubic surface $x^2y + xy^2 + z^2t + zt^2 = 0$ has $q^2 + 7q + 1 = 45$ rational points (i.e. $a = 6$), and that the automorphism group of the surface is the unique subgroup of index 2 of Weyl($E_6$), i.e. is isomorphic to the simple group $\mathbf{SU}_4(\mathbf{F}_4)$, cf. [ATLAS, p.26].
[Hint. If $x \in \mathbf{F}_4$, then $x^2$ is the $\mathbf{F}_2$-conjugate $\overline{x}$ of $x$, so that the equation can be rewritten as $\overline{x}y + x\overline{y} + \overline{z}t + z\overline{t} = 0$, and it is invariant by the unitary group $\mathbf{SU}_4(\mathbf{F}_4)$.]

c) If $a = -3$, show that $N_V(q^e) = (q^e - 1)^2$ if $e \equiv 1, 2 \pmod 3$ and $N_V(q^e) = q^{2e} + 7q^e + 1$ if $e \equiv 0 \pmod 3$.
[Hint. Use the fact that the elements of Weyl($E_6$) of trace $-3$ have order 3, cf. [ATLAS, p.27] and [Do 07, §10.3.3].]

d) (T. Ekedahl and T. Shioda) If $q \equiv 1 \pmod 3$, show that the equation

$$x^3 + y^3 + z^3 + \lambda t^3 = 0,$$

where $\lambda \in \mathbf{F}_q$ is not a cube, defines a smooth cubic surface with $a = -3$.
[Hint. Use Galois descent from the Fermat cubic $x^3 + y^3 + z^3 + t^3 = 0$.]

CHAPTER 3

# THE CHEBOTAREV DENSITY THEOREM FOR A NUMBER FIELD

We limit ourselves to the standard case where the ground field is a number field. For the higher dimensional case, see §9.3.

## 3.1. The prime number theorem for a number field

### 3.1.1. The prime counting function $\pi_K(x)$

Let $K$ be a number field, i.e. a finite extension of $\mathbf{Q}$, and let $O_K$ be the ring of integers of $K$. Let $V_K$ the set of non-archimedean places of $K$. The elements of $V_K$ correspond to the maximal ideals of $O_K$ ; one may identify $V_K$ with $\operatorname{Max} O_K$. We denote by $\mathfrak{p}_v$ the maximal ideal corresponding to $v \in V_K$, and by $|v|$ the norm of $\mathfrak{p}_v$ (also called the norm of $v$), i.e. the number of elements of the residue field $\kappa(v) = O_K/\mathfrak{p}_v$. If $x$ is a real number, we put :

$$\pi_K(x) = \text{ number of } v \in V_K \text{ with } |v| \leqslant x.$$

When $K = \mathbf{Q}$, $V_K$ is the set of prime numbers, and we have

$$\pi_K(x) = \pi(x) = \text{ number of primes } p \text{ with } p \leqslant x.$$

### 3.1.2. The prime number theorem

The theorem says that $\pi_K(x)$ is asymptotically equal to $\pi(x)$, i.e. that $\pi_K(x) \sim x/\log x$ for $x \to \infty$ [1]. More precisely :

**Theorem 3.1.** *There exists* $c > 0$ *such that*

$$|\pi_K(x) - \operatorname{Li}(x)| \ll x \exp(-c\sqrt{\log x}\,) \text{ for } x \geqslant 2.$$

Recall what the $\ll$ notation means : let $A(x)$ and $B(x)$ be two complex functions defined on the same set $\Sigma$, with $B(x)$ real $\geqslant 0$ for all $x \in \Sigma$. One writes :

$$A(x) \ll B(x) \quad \text{for } x \in \Sigma$$

if there exists a real number $C > 0$ such that $|A(x)| \leqslant C.B(x)$ for all $x \in \Sigma$. In most cases, $\Sigma$ is the set of real numbers $x$ larger than some $x_0$ ; if $x_0$ is not specified, then $A(x) \ll B(x)$ is equivalent to $A(x) = O(B(x))$.

[1] Recall that $f(x) \sim g(x)$ means that $\lim_{x\to\infty} f(x)/g(x) = 1$.

Recall also that $\mathrm{Li}(x)$ is the logarithmic integral of $x$, i.e. $\int_2^x dt/\log t$. For every $m > 1$, one has :

$$\mathrm{Li}(x) = \frac{x}{\log x}\left(1 + \frac{1!}{\log x} + \frac{2!}{(\log x)^2} + \cdots + \frac{m!}{(\log x)^m} + O\left(\frac{1}{(\log x)^{m+1}}\right)\right).$$

In particular, $\mathrm{Li}(x) \sim x/\log x$ for $x \to \infty$. Theorem 3.1 implies :

$$\pi_K(x) = x/\log x + x/(\log x)^2 + O(x/(\log x)^3) \quad \text{for } x \to \infty.$$

Note that the expression " logarithmic integral of $x$ " is sometimes used for the slightly different function $\mathrm{li}(x) = \int_0^x dt/\log t$, where the improper integral $\int_0^x$ is defined as $\lim_{\varepsilon \to 0}(\int_0^{1-\varepsilon} + \int_{1+\varepsilon}^x)$. We have $\mathrm{Li}(x) = \mathrm{li}(x) + \mathrm{li}(2) = \mathrm{li}(x) + 1.04516...$; hence most asymptotic statements involving $\mathrm{Li}(x)$ remain true for $\mathrm{li}(x)$. [Note also that some authors use the notation $\mathrm{li}(x)$ for our $\mathrm{Li}(x)$, and vice-versa; there is no universal convention.]

### 3.1.3. Density

Let $P$ be a subset of $\Sigma_K$. For every real number $x$, let $\pi_P(x)$ = number of $v \in P$ with $|v| \leqslant x$. The *upper density* of $P$ is defined by :

$$\text{upper-dens}(P) = \limsup \pi_P(x)/\pi_K(x) \quad \text{for } x \to \infty.$$

The *lower density* of $P$ is defined similarly :

$$\text{lower-dens}(P) = \liminf \pi_P(x)/\pi_K(x) \quad \text{for } x \to \infty.$$

One has lower-dens$(P) \leqslant$ upper-dens$(P)$. When these numbers coincide, they are called the *density* of $P$. Hence, $P$ has density $\lambda$ if and only if :[2]

$$\pi_P(x) = \lambda x/\log x + o(x/\log x) \quad \text{for } x \to \infty.$$

When $K = \mathbf{Q}$, one recovers the usual notion of "natural density" of a set of prime numbers.

*Exercises.*

1) Let $P$ and $Q$ be two subsets of $V_K$. Assume that :

a) $P$ has a density that is $> 0$;

b) upper-dens$(Q) = 1$.

Show that upper-dens$(P \cap Q)$ = dens$(P)$; in particular, $P \cap Q$ is infinite.

---

[2] Recall that " $f(x) = o(g(x))$ for $x \to \infty$ " means that $\lim_{x\to\infty} f(x)/g(x) = 0$.

2) Let $P_1 = \{11, 13, 17, 19, 101, 103, ...\}$ be the set of prime numbers whose first digit in decimal notation is 1.

a) Show that every subset of $P_1$ that has a density has density 0.

b) Show that upper-dens$(P_1) = 5/9$ and lower-dens$(P_1) = 1/9$.

[Hint. Use the prime number theorem, together with the estimate :
$$\sum_{m=1}^{m=n} \frac{10^m}{m} = \frac{10^{n+1}}{9n} + O(\frac{10^n}{n^2}) \text{ for } n \to \infty.]$$

3) Let $P$ and $Q$ be two subsets of $V_K$ which both have a density.

a) Show that

$$\mathrm{d}^-(P \cup Q) + \mathrm{d}^-(P \cap Q) \leqslant \mathrm{d}(P) + \mathrm{d}(Q) \leqslant \mathrm{d}^+(P \cup Q) + \mathrm{d}^+(P \cap Q),$$

where $\mathrm{d}, \mathrm{d}^+$ and $\mathrm{d}^-$ are abbreviations for dens, upper-dens and lower-dens. If $P \cap Q$ and $P \cup Q$ have a density, this implies that $\mathrm{d}(P) + \mathrm{d}(Q) = \mathrm{d}(P \cup Q) + \mathrm{d}(P \cap Q)$.

b) Give an example where neither $P \cap Q$ nor $P \cup Q$ has a density.

[Hint. Let $P_1$ be the set defined in the previous exercise. Choose for $P$ the set of prime numbers that are $\equiv 1 \pmod 3$. Choose for $Q$ the set of primes that are $\equiv 1 \pmod 3$ if they belong to $P_1$, and that are $\equiv -1 \pmod 3$ if not.]

## 3.2. Chebotarev theorem

### 3.2.1. Decomposition group, inertia group, Frobenius

Let $K$ be as in §3.1, and let $E$ be a finite Galois extension of $K$. Let $G$ be the Galois group of $E/K$. The group $G$ acts on the set $V_E$ of the non-archimedean places of $E$, the quotient being $V_K$.

Let $v \in V_K$ and choose $w \in V_E$ lying above $v$. Let $D_w$ be the decomposition group of $w$ in $G = \mathrm{Gal}(E/K)$, i.e. the subgroup of $G$ fixing $w$. Let $I_w$ be the inertia subgroup of $D_w$, i.e. the subgroup made up of the elements $g \in D_w$ that act trivially on the residue field $\kappa(w)$. We have $D_w/I_w \simeq \mathrm{Gal}(\kappa(w)/\kappa(v))$. Let $\sigma_{w/v}$ be the canonical generator of $D_w/I_w$, i.e. the automorphism $x \mapsto x^{|v|}$ of $\kappa(w)$.

Note that $I_w$ is almost always 1 : it is non-trivial if and only if $w$ is ramified over $v$, i.e. if $\mathfrak{p}_v$ divides the discriminant of $E/K$. When $v$ is unramified, $\sigma_{w/v}$ can be viewed as an element of $D_w$ ; it is the *Frobenius* element associated with the pair $(w, v)$[3]. Its conjugacy class in $G$ only depends on $v$ ; we shall denote this class (or any element of it) by $\sigma_v$. When $K = \mathbf{Q}$, we may identify $v$ with the prime number $p = |v|$, and we then write $\sigma_p$ instead of $\sigma_v$.

[3]It will later be called the "arithmetic Frobenius", cf. §4.4 ; its inverse will then be called the "geometric Frobenius".

### 3.2.2. Statement of the theorem – qualitative form

Let $C$ be a subset of $G$ stable under inner automorphisms (i.e. a union of conjugacy classes). Let $V_{K,C} = \{v \in V_K : v \text{ is unramified and } \sigma_v \in C\}$.

**Theorem 3.2.** *The set $V_{K,C}$ has a density; that density is equal to $|C|/|G|$.*

*Remark.* Let $\mathrm{Cl}\,G$ be the set of conjugacy classes of $G$, and let us put on it the measure $\mu$ such that a class $C$ has measure $|C|/|G|$. We may view Theorem 3.2 as an *equidistribution theorem in* $\mathrm{Cl}\,G$ with respect to $\mu$. More precisely, let us order the elements $v$ of $V_K$ in such a way that $v \mapsto |v|$ is increasing[4]. Then the Frobenius classes $\sigma_v$ are equidistributed in $\mathrm{Cl}\,G$ with respect to $\mu$.

One of the most useful consequences of Theorem 3.2 is :

**Corollary 3.3.** *If $C \neq \varnothing$, then $V_{K,C}$ is infinite.*

*Exercise.* Refine Corollary 3.3 by proving that $V_{K,C}$ intersects every subset of $V_K$ whose upper density is 1.
[Hint. Use Exerc.2 of §3.1.3.]

### 3.2.3. Statement of the theorem - quantitative form

If $x$ is a real number $\geqslant 2$, let $\pi_C(x)$ be the number of $v \in V_{K,C}$ with $|v| \leqslant x$. Theorem 3.2 can be refined as follows :

**Theorem 3.4.** (Artin-Chebotarev, cf. [Ar 23], [Ch 25]) *There exists a number $c > 0$ such that*

$$|\pi_C(x) - \frac{|C|}{|G|}\mathrm{Li}(x)| = O(x \exp(-c\sqrt{\log x})) \qquad \text{for } x \to \infty.$$

*If one assumes that the non-trivial zeros of the zeta function of $E$ are on the line $\mathrm{Re}(s) = \frac{1}{2}$, the right hand side can be replaced by $O(x^{\frac{1}{2}} \log x)$.*

For the history of the theorem, see [LS 96].

For a detailed proof, with an effective (or at least theoretically effective) error term, see [LO 77]. For instance, the number " $c$ " of Theorem 3.4 can be taken as $c_0\, n_E^{-\frac{1}{2}}$, where $c_0$ is an effectively computable absolute constant, and $n_E = [E/\mathbf{Q}]$ ; under GRH, one has

$$|\pi_C(x) - \frac{|C|}{|G|}\mathrm{Li}(x)| \leqslant c_1 \frac{|C|}{|G|} x^{\frac{1}{2}} (\log |d_E| + n_E \log x) \qquad \text{for every } x \geqslant 2,$$

where $c_1$ is another effectively computable absolute constant, and $d_E$ is the discriminant of the field $E$. Such explicit error terms are needed when $K$ is fixed but $E$ varies, see e.g. [Se 81].

[4] We follow Bourbaki's conventions : a function $f$ is said to be *increasing* (instead of "non-decreasing") if $x \leqslant y$ implies $f(x) \leqslant f(y)$.

### 3.2.4. Higher moments

We shall later need a variant of Theorem 3.4, where one counts the elements $v$ of $V_{K,C}$ with a weight equal to $|v|^m$ for a given exponent $m$. In order to state the result, let us call $\varepsilon_o(x)$ the function occurring in the error term, namely

$$\varepsilon_o(x) = x\exp(-c\sqrt{\log x}).$$

**Theorem 3.5.** *For $x$ and $m$ real, define $A_m(C,x) = \sum_{|v|\leqslant x,\, v\in V_{K,C}} |v|^m$. If $m \geqslant 0$, we have :*

$$A_m(C,x) = \frac{|C|}{|G|}\mathrm{Li}(x^{m+1}) + O(x^m\varepsilon_o(x)) \quad \textit{for } x \to \infty.$$

*In particular, if $C \neq \varnothing$, then $A_m(C,x) \sim \frac{|C|}{|G|}\cdot\frac{1}{m+1}x^{m+1}/\log x \quad$ for $x \to \infty$.*

*First Proof* (*analytic number theory style*). This follows from Theorem 3.4 by integration by parts.

*Second Proof.* Write $\alpha = \frac{|C|}{|G|}$ for short. For $m = 0$, Theorem 3.4 tells us that $A_0(C,x) = \alpha\mathrm{Li}(x) + \varepsilon(x)$ with $\varepsilon(x) \ll \varepsilon_o(x)$. If $x \geqslant 2$ we have

$$(*) \quad A_m(C,x) = \alpha\mathrm{Li}(x^{m+1}) - \alpha\mathrm{Li}(2^{m+1}) + x^m\varepsilon(x) - \int_2^x mt^{m-1}\varepsilon(t)dt.$$

Indeed, both sides of this equation have the following properties, which are strong enough to imply that they are equal :

• they are differentiable at every point $x$, except possibly when $x \in \mathbf{N}$, in which case they are right-continuous and they jump by the same amount, namely $x^m a(x)$ where $a(x)$ is the number of unramified $v$ with $\sigma_v \in C$ and $|v| = x$;

• their values at $x = 2$ are the same, namely $2^m a(2)$;

• their derivatives at every $x \notin \mathbf{N}$ are equal to $0$; this is clear for the left side; the derivative of the right side is

$$\alpha x^m/\log x + mx^{m-1}\varepsilon(x) + x^m\varepsilon'(x) - mx^{m-1}\varepsilon(x) = x^m(\alpha/\log x + \varepsilon'(x)),$$

which is 0 because $\varepsilon'(x) = -\alpha\, d\mathrm{Li}(x)/dx = -\alpha\,/\log x$.

Formula (*) implies Theorem 3.5, since $x^m\varepsilon(x) \ll x^m\varepsilon_o(x)$ and :

$$\int_2^x mt^{m-1}\varepsilon(t)dt \ll \int_2^x t^{m-1}\varepsilon_o(t)dt \ll x^m\varepsilon_o(x),$$

the last inequality being due to the fact that $\varepsilon_o(x)$ is increasing when $x$ is large enough.

*Remark.* The proof applies to every $\varepsilon_o \gg \varepsilon$ which is increasing in a neighborhood of $\infty$. In particular, it applies to $\varepsilon_o(x) = x^{\frac{1}{2}} \log x$ if GRH is true for the field $E$.

*Variant.* It is often convenient to express Theorem 3.5 in terms of class functions on $G$. If $f$ is such a function (with values in $\mathbf{C}$, say), define

$$A_m(f,x) = \sum_{|v| \leqslant x} |v|^m f(\sigma_v),$$

where the sum extends over the $v$ that are unramified in $E/K$ (so that $\sigma_v$ is well defined) and have norm $\leqslant x$. With this notation, Theorem 3.5 can be reformulated as :

**Theorem 3.6.** *Assume* $m \geqslant 0$. *Then* :

$$A_m(f,x) \;=\; <f,1>_G \operatorname{Li}(x^{m+1}) + O(x^m \varepsilon_o(x)) \quad \textit{for} \quad x \to \infty,$$

*where* $<f,1>_G = \frac{1}{|G|}\sum_{g \in G} f(g)$ *is the mean value of* $f$ *on* $G$.

Indeed, this is the same statement as Theorem 3.5 when $f$ is the characteristic function of a subset $C$ of $G$ that is stable under conjugation; the general case follows by linearity.

*Exercise.* With the same notation as in Theorem 3.5, show that $A_{-1}(C,x) - \frac{|C|}{|G|} \log\log x$ has a limit for $x \to \infty$.

[Hint. Use partial summation.]

## 3.3. Frobenian functions and frobenian sets

### 3.3.1. $S$-frobenian functions and $S$-frobenian sets

Let $S$ be a finite subset of $V_K$, and let $\Omega$ be a set (with the discrete topology). Consider a map $f : V_K - S \to \Omega$. We say that $f$ is *$S$-frobenian* if there exists a finite Galois extension $E/K$, unramified outside $S$, and a map $\varphi : G \to \Omega$, where $G = \operatorname{Gal}(E/K)$, such that :

a) $\varphi$ is invariant under conjugation (i.e. $\varphi$ factors through $G \to \operatorname{Cl} G$).

b) $f(v) = \varphi(\sigma_v)$ for all $v \in V_K - S$.

[Note that $\varphi(\sigma_v)$ makes sense because of condition a).]

A subset $\Sigma$ of $V_K - S$ is said to be *$S$-frobenian* if its characteristic function is $S$-frobenian. This means that there exists a Galois extension $E/K$ as above, and a subset $C$ of its Galois group $G$, stable under conjugation, such that $v \in \Sigma \iff \sigma_v \in C$; in that case, Theorem 3.4 shows that $\Sigma$ has a density, which is equal to $|C|/|G|$.

Here is an alternative form of the definition. Let $\overline{K}$ be an algebraic closure of $K$ and let $K_S$ be the maximal subextension of $\overline{K}$ that is unramified outside $S$. Let $\Gamma_S$ be the Galois group of $K_S$ over $K$ ; let $\mathrm{Cl}\,\Gamma_S$ be the set of its conjugacy classes, with its natural profinite topology (quotient of that of $\Gamma_S$). We have :

**Proposition 3.7.**

a) *A function* $f : V_K - S \to \Omega$ *is* $S$*-frobenian if and only if there exists a continuous map* $\varphi : \mathrm{Cl}\,\Gamma_S \to \Omega$ *such that* $f$ *is equal to the composition*

$$V_K - S \to \mathrm{Cl}\,\Gamma_S \xrightarrow{\varphi} \Omega,$$

*where the map on the left is* $v \mapsto \sigma_v$.
*When this is the case,* $\varphi$ *is unique; its image is a finite subset of* $\Omega$, *that is equal to the image of* $f$.

b) *A subset* $P$ *of* $V_K - S$ *is* $S$*-frobenian if and only if there exists a open and closed subset* $U$ *of* $\mathrm{Cl}\,\Gamma_S$ *such that* $v \in P \iff \sigma_v \in U$. *When is the case,* $U$ *is unique; it is the closure of the set of* $\sigma_v$, *for* $v \in P$. *The set* $P$ *has a density that is equal to the Haar measure*[5] *of* $U$.

*Proof of* a). Since $\Omega$ is discrete, the continuity of $\varphi$ means that $\varphi$ is locally constant, i.e. factors through $\mathrm{Cl}\,\Gamma_S/N$, where $N$ is an open normal subgroup of $\Gamma_S$. Hence, the first assertion is just a restatement of the definition. The uniqueness of $\varphi$ follows from Chebotarev theorem, since the $\sigma_v$'s are dense in $\mathrm{Cl}\,\Gamma_S$ ; the same argument shows that the images of $f$ and $\varphi$ are the same, and that they are finite.

The proof of b) is similar; the assertion about the density is merely a reformulation of Theorem 3.2.

*Notation.* If $f$ is $S$-frobenian, the corresponding map $\varphi$ will be denoted by $\varphi_f$ ; we shall view it indifferently as a map of $\Gamma_S$ or of $\mathrm{Cl}\,\Gamma_S$ into $\Omega$. There exists a maximal open normal subgroup $N$ of $\Gamma_S$ such that $\varphi_f$ is constant mod $N$. Equivalently, there exists a minimal finite Galois extension $E_f/K$, unramified outside $S$, such that $\varphi_f$ factors through $\mathrm{Gal}(E_f/K)$ ; we shall say that $E_f$ is associated with $f$.

*Remark.* Part b) of Proposition 3.7 gives natural bijections between :

a) $S$-frobenian subsets of $V_K - S$;

b) open and closed subsets of $\Gamma_S$ stable under conjugation;

c) open and closed subsets of $\mathrm{Cl}\,\Gamma_S$.

[5] We always use the *normalized* Haar measure, i.e. the Haar measure with total mass 1, see §5.2.1. It is a measure on $\Gamma_S$ ; its image by $\Gamma_S \to \mathrm{Cl}\,\Gamma_S$ is a measure on $\mathrm{Cl}\,\Gamma_S$, that we also call the "Haar measure". If $A$ is a measurable subset of $\Gamma_S$, that is stable under conjugation, and if $\mathrm{cl}(A)$ is its image in $\mathrm{Cl}\,\Gamma_S$, the Haar measures of $A$ and of $\mathrm{cl}(A)$ are the same.

These bijections are compatible with taking finite unions, finite intersections, and complements. In particular, the intersection of two $S$-frobenian sets is $S$-frobenian.

**Proposition 3.8.**

i) *If an $S$-frobenian set is non-empty, its density is* $> 0$; *in particular, it is infinite.*

ii) *Let $P$ and $P'$ be two $S$-frobenian subsets of $V_K - S$; assume that there are two sets $Q$ and $Q'$, of density* 0, *such that $P \cup Q = P' \cup Q'$. Then $P = P'$.*

*Proof.*

i) If $P$ is $S$-frobenian and non-empty, the corresponding open and closed subset of $\Gamma_S$ is non empty, hence its Haar measure is $> 0$. This shows that $\mathrm{dens}(P) > 0$.

ii) The set $E = P \cup P' - P \cap P'$ is contained in $Q \cup Q'$, hence has density 0. Since $E$ is $S$-frobenian, this implies $E = \varnothing$.

### 3.3.2. Frobenian sets and frobenian functions

There are cases where one does not want to specify a set $S$ as we did in the previous section. This leads to the following definition :

A subset $P$ of $V_K$ is called *frobenian* if there exists a finite set $S$ such that $P - S \cap P$ is $S$-frobenian. Such a set defines an open and closed subset $U_P$ of $\Gamma_K = \mathrm{Gal}(\overline{K}/K)$, closed under conjugation, which one may define either as :

a) the closure of the union of the conjugacy classes of the $\sigma_v$, for $v \in P - S$ (this is independent of the choice of $S$);

b) the inverse image by $\Gamma_K \to \Gamma_S \to \mathrm{Cl}\,\Gamma_S$ of the set denoted by $U$ in Proposition 3.7.b).

Two frobenian sets $P$ and $P'$ are called *almost equal* if they only differ by a finite set, i.e. if there exists a finite set $S$ such that $P - P \cap S = P' - P' \cap S$. This is equivalent to $U_P = U_{P'}$. One thus gets a natural bijection between :

a) frobenian subsets of $V_K$, up to the equivalence relation defined by "almost equality";

b) open and closed subsets of $\Gamma_K$ that are stable under conjugation;

c) open and closed subsets of $\mathrm{Cl}\,\Gamma_K$.

The properties of $S$-frobenian sets proved in the previous section imply :

**Proposition 3.9.** i) *If a frobenian set is infinite, its density is* $> 0$.

ii) *Let $P$ and $P'$ be two frobenian subsets of $V_K$; assume that there are two sets $Q$ and $Q'$, of density* 0, *such that $P \cup Q = P' \cup Q'$. Then $P$ and $P'$ are almost equal.*

Similarly, if $S_0$ is a finite subset of $V_K$, a map $f : V_K - S_0 \to \Omega$ is called *frobenian* if there exists a finite set $S$ containing $S_0$ such that the restriction of $f$ to $V_K - S$ is $S$-frobenian. The fibers of such a map are frobenian subsets of $V_K$.

*Exercises.*

1) Let Dens (resp. Frob) be the set of all subsets of $V_K$ that have a density (resp. are frobenian). Show that $|\mathrm{Frob}| = \aleph_0$ and $|\mathrm{Dens}| = 2^{\aleph_0}$. [Hence Frob is much smaller than Dens.]

2) Let $Q$ be a frobenian subset of $V_K$ and let $\alpha$ be its density. Show that there exists $c > 0$ such that $\prod_{v \in Q, |v| \leqslant x}(1 - \frac{1}{|v|}) \sim \frac{c}{(\log x)^\alpha}$ for $x \to \infty$.
[Hint. Use the exercise in §3.2.4.]

### 3.3.3. Basic properties of $S$-frobenian functions

Let $f : V_K - S \to \Omega$ be an $S$-frobenian function.

3.3.3.1. If $\omega$ is an element of $\Omega$, the set $f^{-1}(\omega)$ is $S$-frobenian, hence is either empty or has a density that is $> 0$.

If $f' : V_K - S \to \Omega$ is another $S$-frobenian function, the set of $v \in V_K - S$ such that $f(v) = f'(v)$ is either empty, or has a density that is $> 0$; this follows from the fact that $(f, f') : V_K - S \to \Omega \times \Omega$ is $S$-frobenian.

Similarly, the fibers of a frobenian function are either finite, or of density $> 0$.

3.3.3.2. *Value at* 1 *and at* $-1$. One can define the *value of $f$ at* 1 as being $\varphi_f(1)$, where " 1 " means the identity element of $\Gamma_S$ (or of $\mathrm{Gal}(E_f/K)$, it amounts to the same). Note that the element " $f(1)$ " so defined belongs to the image of $f$, since it belongs to the image of $\varphi_f$; hence *the set of $v \in V_K - S$ with $f(v) = f(1)$ is $S$-frobenian of density* $> 0$.

Similarly, if $\iota : K \to \mathbf{R}$ is a real embedding of $K$, and $c_\iota$ is the corresponding complex conjugation (viewed as an element of $\mathrm{Cl}\,\Gamma_S$), we can define the *value of $f$ at* $-1_\iota$ as $\varphi_f(c_\iota)$; here again, the set of $v \in V_K - S$ with $f(v) = f(-1_\iota)$ is $S$-frobenian of density $> 0$. When $K = \mathbf{Q}$, we shall write[6] $f(-1)$ instead of $f(-1_\iota)$.

3.3.3.3. $\Psi^e$-*transforms.* Let $e$ be an integer. If $\varphi$ is any function defined on a group $G$, let us denote[7] by $\Psi^e\varphi$ the function $g \mapsto \varphi(g^e)$. This applies in particular to $G = \Gamma_S$ and $\varphi = \varphi_f$. Since $\Psi^e\varphi_f$ is a locally constant class function on $G - S$, it defines an $S$-frobenian map of $V_K - S$ into $\Omega$,

[6]These strange notations, which amount to speaking of "the prime 1" and "the prime $-1$", lead to very reasonable-looking formulae, as we shall see in §3.4.1.1 and §6.1.2. Note that " $-1$ " was already advertised by Conway in [Co 97 ] as a convenient index for the real place at infinity.

[7]The $\Psi$ notation comes from the Adams operations for group representations and K-theory, see §5.1.1 or [Se 78, §9.1, exerc.].

which we shall denote by $\Psi^e f$. By definition, we have $\Psi^e f(v) = f(\sigma_v^e)$ for every $v \in V_K - S$, where $\sigma_v^e$ denotes the $e$-th power of the Frobenius element $\sigma_v$. We have $\Psi^e f(1) = f(1), \Psi^e f(-1_\iota) = f(1)$ if $e$ is even, and $\Psi^e f(-1_\iota) = f(-1_\iota)$ if $e$ is odd.

3.3.3.4. *Base change.* Let $K'$ be a finite extension of $K$ contained in $K_S$, hence unramified outside $S$, and let $G'_S = \mathrm{Gal}(K_S/K')$ be the corresponding subgroup of $\Gamma_S$.

Let $f : V_K - S \to \Omega$ be an $S$-frobenian map, and let $S'$ be the inverse image of $S$ in $V_{K'}$. There is a unique $S'$-frobenian map $f' : V_{K'} - S' \to \Omega$ such that $\varphi_{f'} : G'_S \to \Omega$ is the restriction of $\varphi_f$ to $G'_S$. If $v' \in V_{K'} - S'$ has image $v$ in $V_K - S$, and if $e$ is the corresponding residue degree, we have :

$$f'(v') = \varphi_{f'}(\sigma_{v'}) = \varphi_f(\sigma_v^e) = \Psi^e f(v).$$

Hence $f'$ can be computed, provided one knows $\Psi^e f$ for $e \leqslant [K' : K]$.

*Warning.* The map $f'$ should not be confused with the more natural looking composition $V_{K'} - S' \to V_K - S \to \Omega$, which is not frobenian in general, even when $[K' : K] = 2$, cf. Exercise 1 below.

3.3.3.5. *Mean value.* Suppose $\Omega$ is a $\mathbf{Q}$-vector space. We may then define the *mean value* mean$(f)$ of $f$ as that of $\varphi_f$, namely

$$\mathrm{mean}(f) \;= <\varphi_f, 1> = \frac{1}{|G|} \sum_{g \in G} \varphi_f(g),$$

where $G$ is any finite quotient of $\Gamma_S$ through which $\varphi_f$ can be factored. If $\Omega$ is $\mathbf{C}$, or more generally a finite-dimensional vector space over $\mathbf{R}$, Theorem 3.4 implies :

$$\sum_{|v| \leqslant x} f(v) = \mathrm{mean}(f)\,\mathrm{Li}(x) + O(\varepsilon_o(x)) \;\text{ for } x \to \infty,$$

where $\varepsilon_o(x)$ is as in §3.2.4, and the summation is restricted to $v \notin S$. In particular, we have $\mathrm{mean}(f) = \lim_{x\to\infty} \frac{\sum_{|v|\leqslant x} f(v)}{\pi_K(x)}$.

Similarly, Theorem 3.5 implies :

$$\sum_{|v| \leqslant x} f(v)|v|^m = \mathrm{mean}(f)\,\mathrm{Li}(x^{m+1}) + O(x^m \varepsilon_o(x)) \;\text{ for } x \to \infty,$$

for every $m \geqslant 0$.

*Exercises.* We keep the notation of §3.3.3.3 and we denote by $\pi$ the natural projection $V_{K'} - S' \to V_K - S$.

1) Suppose $[K' : K] = 2$. Let $A$ be the subset of $V_K - S$ made up of the $v \in V_K$ that split completely in $K'$. Show that $A$ is frobenian, but $\pi^{-1}(A)$ is not.

[Hint. Observe that $\pi^{-1}(A)$ has density 1, but that its complement is infinite.]

Use $A$ to construct an $S$-frobenian function $f : V_K - S \to \mathbf{Z}/2\mathbf{Z}$ such that $f \circ \pi$ is not frobenian.

2) Let $E$ be a Galois extension of $K$ containing $K'$ and unramified outside $S$; put $G = \mathrm{Gal}(E/K)$ and $G' = \mathrm{Gal}(E/K')$ so that we have $G' \subset G$. Let $\psi$ be a class function on $G'$ with values in $\mathbf{R}$. Define a function $\varphi$ on $G$ by the formula

$$\varphi(g) = \sup_{x \in G} \psi(x g^{e(x,g)} x^{-1}),$$

where $e(x, g)$ is the smallest integer $n > 0$ such that $x g^n x^{-1} \in G'$.

a) Show that $\varphi$ is a class function on $G$.

b) Let $f_\varphi : V_K - S \to \mathbf{R}$ and $f_\psi : V_{K'} - S' \to \mathbf{R}$ be the frobenian functions associated with $\varphi$ and $\psi$. Show that

$$f_\varphi(v) = \sup_{v' \to v} f_\psi(v').$$

c) Let $B$ be an $S'$-frobenian subset of $V_{K'} - S'$. Show that $A = \pi(B)$ is an $S$-frobenian subset of $V_K - S$.

[Hint. Apply b) to a sufficient large extension of $K'$ contained in $K_S$, and choose $\psi$ such that $f_\psi$ is the characteristic function of $B$. Then $f_\varphi$ is the characteristic function of $A$.]

## 3.4. Examples of $S$-frobenian functions and $S$-frobenian sets

In this section, the ground field $K$ is $\mathbf{Q}$, so that $V_K$ is the set $P$ of all prime numbers.

### 3.4.1. Dirichlet examples

The following two examples are essentially due to Dirichlet ([Di 39]) :

3.4.1.1. *Arithmetic progressions.* Let $m$ be an integer $> 0$ and let $S$ be the set of the prime divisors of $m$. Let $f : P - S \to (\mathbf{Z}/m\mathbf{Z})^\times$ be the function $p \mapsto p \bmod m$. Then $f$ is $S$-frobenian, the relevant field extension being the $m$-th cyclotomic field. One has

$$f(1) = 1 \quad \text{and} \quad f(-1) = -1,$$

where $f(1)$ and $f(-1)$ are the elements of $(\mathbf{Z}/m\mathbf{Z})^\times$ associated with $f$ as explained in §3.3.2.2.

3.4.1.2. *Binary quadratic forms.* Let $B(x,y) = ax^2 + bxy + cy^2$ be a binary quadratic form, with integral coefficients, whose discriminant $d = b^2 - 4ac$ is not a square. Let $S$ be the set of the prime divisors of $d$. Let $P_B$ be the set of primes $p \notin S$ that can be represented by $B$, i.e. are of the form $p = B(x,y)$ for some $x, y \in \mathbf{Z}$. Then $P_B$ is $S$-frobenian. It is not empty if the obvious necessary conditions are met : $(a,b,c) = 1$ and $a > 0$ if $d < 0$; the density of $P_B$ is then $1/h^+(d)$, except if $B$ is *ambiguous* , i.e. invariant by an element of $\mathbf{GL}_2(\mathbf{Z})$ of determinant $-1$, in which case the density is $1/2h^+(d)$. [Here $h^+(d)$ is the narrow class number, cf. [Co 93, 5.2.7] ; when $d$ is $> 0$ it may differ by a factor 2 from the usual class number $h(d)$.]

The proofs of these statements rely on the standard dictionary between binary quadratic forms and invertible ideals in quadratic rings (see [Co 93, §5.2]). The relevant Galois extensions are the abelian extensions of $\mathbf{Q}(\sqrt{d})$ known as ring class fields, cf. [Cox 89, §9].

*Exercise.* Show that the primes represented by $2x^2 + xy + 9y^2$ have density $1/7$.

### 3.4.2. The map $p \mapsto N_X(p)$

3.4.2.1. *The case where* $\dim X_{/\mathbf{Q}} \leqslant 0$. Let $X$ be a scheme of finite type over $\mathbf{Z}$. Assume that $\dim X_{/\mathbf{Q}} \leqslant 0$, i.e. that $X(\overline{\mathbf{Q}})$ is a finite set. The Galois group $\Gamma_{\mathbf{Q}} = \mathrm{Gal}(\overline{\mathbf{Q}}/\mathbf{Q})$ acts on $X(\overline{\mathbf{Q}})$ via some quotient $G = \mathrm{Gal}(E/\mathbf{Q})$, where $E$ is a finite Galois extension of $\mathbf{Q}$.

**Proposition 3.10.**
a) *The map* $f : P \mapsto \mathbf{Z}$ *defined by* $f(p) = N_X(p)$ *is frobenian.*
b) *The corresponding map* $\varphi_f$ *(cf.* §3.3.1*) is the map* $\varphi : G \to \mathbf{Z}$ *defined by :*

$$\varphi(g) = X(\overline{\mathbf{Q}})^g = \textit{ number of points of } X(\overline{\mathbf{Q}}) \textit{ fixed by } g.$$

c) $\Psi^e f(p) = N_X(p^e)$ *for every large enough* $p$ *and every* $e \geqslant 1$. [For the definition of $\Psi^e f$, see §3.3.3.3.]
d) *One has* $f(1) = |X(\mathbf{C})|$ *and* $f(-1) = |X(\mathbf{R})|$.
e) *The mean value of* $f$ (in the sense of §3.3.3.5) *is* $|X_{/\mathbf{Q}}|$.

*Proof.* Let us suppose first that $X = \mathrm{Spec}\, O_K$, where $K$ is a number field. We then have $N_X(p) =$ number of places $v$ of $K$ with $|v| = p$.

We may choose for $E$ the Galois closure of $K$ ; we have $G = \mathrm{Gal}(E/\mathbf{Q})$ ; let us put $H = \mathrm{Gal}(E/K)$. We may identify $X(\overline{\mathbf{Q}})$ with $G/H$. Let $S$ be the set of prime divisors of $\mathrm{disc}(K)$ (or of $\mathrm{disc}(E)$ – it amounts to the same). If $p \notin S$, let $\sigma_p \in G$ be its Frobenius element (up to conjugation). We have $N_X(p) = \varphi(\sigma_p)$. This shows that the map $p \mapsto N_X(p)$ is

$S$-frobenian, and that $\varphi$ is the associated function. This proves a) and b). Assertion c) follows from the easily proved formula $N_X(p^e) = \varphi(\sigma_p^e)$. Since $\varphi(1) = |G/H| = [K/\mathbf{Q}] = |X(\mathbf{C})|$, this implies the assertion about $f(1)$; a similar method works for $f(-1)$. As for the mean value of $f$, it is by definition the mean value of the function $\varphi$, that is equal to 1 by Burnside's lemma (see e.g. [Se 02, §2.1]). This proves Proposition 3.10 in the case $X = \operatorname{Spec} O_K$. The general case follows by doing the following operations on $X$ :

- making it reduced; this does not change any $N_X(p^e)$;
- making it normal; this changes $N_X(p^e)$ for only finitely many $p$;
- decomposing it into irreducible components; each component is then isomorphic to $X' = \operatorname{Spec} O_K - \Sigma$, where $K$ is a number field and $\Sigma$ is a closed finite subset of $\operatorname{Spec} O_K$; one then has $N_X(p^e) = N_{X'}(p^e)$ for all large enough primes $p$.

**Corollary 3.11.** *If $X$ is a scheme of finite type over $\mathbf{Z}$ such that $X_{/\mathbf{Q}} \neq \varnothing$, there are infinitely many $p$ with $N_X(p) > 0$.*

*Proof.* Choose a closed point $x$ of $X_{/\mathbf{Q}}$; its closure $X_x$ in $X$ is a subscheme of $X$ to which one applies part c) of Proposition 3.10. Hence there are infinitely many $p$ with $N_{X_x}(p) > 0$, and *a fortiori* $N_X(p) > 0$.

*Remarks.* 1. The hypothesis $X_{/\mathbf{Q}} \neq \varnothing$ is equivalent to $X(\overline{\mathbf{Q}}) \neq \varnothing$ and to $X(\mathbf{C}) \neq \varnothing$.

2. By a theorem of Ax and van den Dries ([Ax 67], [Dr 91], see also §7.2.4), the set of $p$ with $N_X(p) \neq 0$ is frobenian; in particular, it has a density, that is $> 0$, as was first proved in [Ax 67].

*Example : Number of roots mod $p$ of a one-variable polynomial.*

Let us take $X = \mathbf{Z}[t]/(H)$, where $H$ is a non-zero element of the polynomial ring $\mathbf{Z}[t]$. Let $a_0 t^n$ be the leading term of $H$ and let $S$ be the set of the prime divisors of $a_0 \operatorname{disc}(H)$. Then :

*The map $p \mapsto N_H(p)$ is $S$-frobenian, and its value at* 1 (*resp. at* $-1$) *is the number of complex* (*resp. real*[8]) *roots of $H$.*

*For every $e \geqslant 1$, the $\Psi^e$-transform of $p \mapsto N_H(p)$ is $p \mapsto N_H(p^e)$.*

3.4.2.2. $N_X(p)$ mod $m$. Let $X$ be a scheme of finite type over $\mathbf{Z}$. The map $p \mapsto N_X(p)$ is not frobenian (unless $\dim X(\mathbf{C}) \leqslant 0$), if only because its image is infinite, but it is *residually frobenian* in the following sense :

[8]Note the following curious corollary : there are infinitely many $p$ such that $H$ has the same number of roots in $\mathbf{F}_p$ as in $\mathbf{R}$.

*For every integer* $m \geqslant 1$, *the map* $f : p \mapsto N_X(p) \pmod{m}$ *is a frobenian map of* $P$ *into* $\mathbf{Z}/m\mathbf{Z}$; *its value at* 1 (resp. *at* $-1$) *is equal to the image in* $\mathbf{Z}/m\mathbf{Z}$ *of the Euler-Poincaré characteristic* $\chi(X(\mathbf{C}))$ (resp. $\chi_c(X(\mathbf{R}))$).

These statements will be proved in §6.1.2; note that they imply Theorem 1.4 of §1.4.

### 3.4.3. The $p$-th coefficient of a modular form

Let us choose a level $N$, a weight $k > 0$, a Dirichlet character $\varepsilon$ mod $N$, and a modular form $\varphi = \sum a_n q^n$ on $\Gamma_0(N)$ of weight $k$ and type $\varepsilon$, cf. e.g. [DS 74, §1]. Assume that the coefficients $a_n$ of $\varphi$ belong to the ring of integers $A$ of some number field. Then *the map* $p \mapsto a_p$ *is residually frobenian*, in a similar sense as in §3.4.2.2. More precisely :

**Theorem 3.12.** *For every integer* $m \geqslant 1$, *the map* $p \mapsto a_p \pmod{mA}$ *is* $S_{mN}$*-frobenian, where* $S_{mN}$ *is the set of primes dividing* $mN$. *Its value at* 1 (*resp. at* $-1$) *is* $2a_1$ (*resp.* 0) $\pmod{mA}$.

**Corollary 3.13.** *The set of* $p$ *with* $a_p \equiv 2a_1 \pmod{m}$ *is* $S_{mN}$*-frobenian and its density is* $> 0$. *The same is true for the set of* $p$ *with* $a_p \equiv 0 \pmod{m}$.

The proof of Theorem 3.12 is by reduction to the case where $m$ is a power of a prime $\ell$, and $\varphi$ is an eigenfunction of the Hecke operators $T_p$ for $p \notin S_{\ell N}$ ; in that case, one uses the fact that there exists a 2-dimensional $\ell$-adic representation $\rho$ of $\Gamma_{S_{\ell N}}$ such that $a_p = a_1 \mathrm{Tr}(\rho(\sigma_p))$ for every $p \notin S_{\ell N}$.

Similarly, if $H(N,k)$ is the ring generated by the Hecke operators $T_p$ for $p$ not dividing $N$, *the map* $p \mapsto T_p \in H(N,k)$ *is residually frobenian*, and its value at 1 (resp. at $-1$) is 2 (resp. 0). In particular, if $\varphi$ is as above, one has $T_p\varphi \equiv 0 \pmod{m}$ for a set of $p$ that is $S_{mN}$-frobenian of density $\delta > 0$. As explained in [Se 76, §4.6], this implies that $\varphi$ is lacunary (mod $m$), i.e., if we denote by $s_m(x)$ the number of $n \leqslant x$ with $a_n \equiv 0 \pmod{m}$, we have $\lim_{x\to\infty} s_m(x)/x = 1$ ; more precisely : $x - s_m(x) = O(x/(\log x)^\delta)$.

*Exercise.* Let $m \geqslant 1$ and $\varphi = \sum a_n q^n$ be as above, and let $e$ be an integer $\geqslant 1$. Show that the map $p \mapsto a_{p^e} \pmod{m}$ is frobenian, that its value at 1 is $(e+1)a_1$ and that its value at $-1$ is 0 if $e$ is odd and is $a_1$ if $e$ is even.

### 3.4.4. Examples of non-frobenian sets of primes

The reader should not think that all reasonably defined sets of primes are frobenian. Here is a counterexample :

Choose $X$ such that $X_{/\mathbf{Q}}$ is a geometrically irreducible smooth projective curve of genus $g > 0$, and let $P_X$ be the set of primes $p$ such that

$N_X(p) > p+1$. It is a consequence of the general Sato-Tate conjecture (cf. Chapter 8) that $P_X$ is not frobenian, but that it has a density, which is a strictly positive rational number $\leqslant \frac{1}{2}$.

When $X_{/\mathbf{Q}}$ is an elliptic curve (i.e. $g=1$), most of this has been proved (see the references given in §8.1.5), the density of $P_X$ being equal to $\frac{1}{4}$ in the CM case and $\frac{1}{2}$ in the non-CM case. In the second case, the non-frobenian behaviour of $P_X$ should be true in the following strong form : if $F$ is any frobenian set of primes, then $F \cap P_X$ has a density equal to $\mathrm{dens}(P_X).\mathrm{dens}(F) = \frac{1}{2}\mathrm{dens}(F)$ ; loosely speaking, the condition $p \in P_X$ is independent of any frobenian condition.

$V(\beta) \geq$ [illegible]. It is a consequence of the general Salem [illegible] (Theorem [illegible]) that [illegible], but [illegible] if [illegible] has all [illegible] strictly [illegible] rational number.

[illegible] with [illegible] (the [illegible]) [illegible] that [illegible] the [illegible] the [illegible] of [illegible] is [illegible] to the [illegible] and $\frac{1}{2}$ in the [illegible] case. In the second case, the [illegible] of [illegible] should be true. In the following we [illegible] [illegible] that [illegible] has [illegible] equal to [illegible] [illegible] independent of any [illegible] conditions.

## CHAPTER 4

# REVIEW OF $\ell$-ADIC COHOMOLOGY

The results summarized in this chapter (except those of §4.6) can be found in the three volumes of SGA relative to étale cohomology : [SGA 4], [SGA $4\frac{1}{2}$] and [SGA 5], together with Deligne's papers [De 74] and [De 80] on Weil's conjectures. For a shorter account (with or without proofs), see e.g. [FK 88], [Ka 94], [Ka 01b] or [Mi 80].

### 4.1. The $\ell$-adic cohomology groups

Let $k$ be a field with separable closure $k_s$ and let $\Gamma_k = \mathrm{Gal}(k_s/k)$.

Let $X$ be a $k$-variety, i.e. a scheme of finite type over $k$, and let $\overline{X} = X_{/k_s}$ be the $k_s$-variety obtained from $X$ by the base change $k \to k_s$. Let us also assume that $X$ is separated. Let $\ell$ be a prime number, distinct from $\mathrm{char}(k)$ (the choice of $\ell$ is usually unimportant, the only exception in the present notes being §6.3.4). For every $i \geqslant 0$, one defines the étale cohomology groups $H^i(\overline{X}, \mathbf{Q}_\ell)$ and $H^i_c(\overline{X}, \mathbf{Q}_\ell)$ ; they are finite dimensional $\mathbf{Q}_\ell$-vector spaces [1], and they vanish when $i > 2\dim(X)$; when $X$ is proper, the two types coincide. We shall mainly use the $H^i_c$, i.e. those with a " $c$ " in index position, which are called the *i-th cohomology groups of* $\overline{X}$ *with proper* [2] *support.* One advantage of these groups is that, if $Y$ is a closed subvariety of $X$ and $U = X - Y$, one has the usual type of exact sequence

$$\cdots \to H^i_c(\overline{X}, \mathbf{Q}_\ell) \to H^i_c(\overline{Y}, \mathbf{Q}_\ell) \to H^{i+1}_c(\overline{U}, \mathbf{Q}_\ell) \to H^{i+1}_c(\overline{X}, \mathbf{Q}_\ell) \to \cdots$$

The action of $\Gamma_k$ on $\overline{X}$ gives by "transport de structure" an action of $\Gamma_k$ on each of the vector spaces $H^i(\overline{X}, \mathbf{Q}_\ell)$ and $H^i_c(\overline{X}, \mathbf{Q}_\ell)$. We thus get $\ell$-adic representations of $\Gamma_k$, and hence $\ell$-adic characters, see Chapters 6 and 7.

*Remark.* The $\ell$-adic cohomology groups are *topological* invariants, in the following sense : they are the same for a scheme $X$ and for the corresponding reduced scheme $X^{\mathrm{red}}$ ; for a more general result (valid for any finite surjective radicial morphism), see [SGA 4, VIII, th.1.1] and [SGA 1, IX, th.4.10].

[1] When $X$ is proper and smooth, the dimension of these spaces does not depend on the choice of $\ell$ ; the same is true if $\mathrm{char}(k) = 0$ because of Artin's comparison theorem, see §4.2 below ; whether this holds in general is a well-known open question.

[2] One also says "with compact support", by analogy with the case $k = \mathbf{C}$ ; this explains the use of the index " $c$ " in the notation.

*Remark on the definitions of* $H^i(\overline{X}, \mathbf{Q}_\ell)$ *and* $H^i_c(\overline{X}, \mathbf{Q}_\ell)$. These definitions will not actually be used in what follows. However, the reader should be aware of the fact that, despite its suggestive notation, $H^i(\overline{X}, \mathbf{Q}_\ell)$ is *not* the $i$-th cohomology group of $\overline{X}$ with coefficients in the constant sheaf $\mathbf{Q}_\ell$. The groups that are genuine sheaf cohomology groups (for the étale topology) are the finite groups $H^i(\overline{X}, \mathbf{Z}/\ell^n\mathbf{Z})$ for $n = 1, 2, \ldots$; the groups $H^i(\overline{X}, \mathbf{Z}_\ell)$ and $H^i(\overline{X}, \mathbf{Q}_\ell)$ are derived from them by the formulae

$$H^i(\overline{X}, \mathbf{Z}_\ell) = \varprojlim H^i(\overline{X}, \mathbf{Z}/\ell^n\mathbf{Z})$$

and

$$H^i(\overline{X}, \mathbf{Q}_\ell) = H^i(\overline{X}, \mathbf{Z}_\ell) \otimes_{\mathbf{Z}_\ell} \mathbf{Q}_\ell = H^i(\overline{X}, \mathbf{Z}_\ell)[1/\ell].$$

As for the $H^i_c(\overline{X}, \mathbf{Z}_\ell)$ and $H^i_c(\overline{X}, \mathbf{Q}_\ell)$, they are defined in a similar way, by means of a compactification of $X$.

A consequence of these somewhat indirect definitions is that most of the basic results (such as those on higher direct images, or base change) have to be proved first for the constant sheaves $\mathbf{Z}/\ell^n\mathbf{Z}$ (or, more generally, for finite constructible sheaves), and then extended to $\mathbf{Z}_\ell$ by using the functor $\varprojlim$, and extended to $\mathbf{Q}_\ell$ by using the functor $\otimes_{\mathbf{Z}_\ell}\mathbf{Q}_\ell$.

## 4.2. Artin's comparison theorem

Suppose $k = \mathbf{R}$ or $\mathbf{C}$, so that $k_s = \overline{k} = \mathbf{C}$. In that case, the $\mathbf{C}$-analytic space $X(\mathbf{C})$ is locally compact for the usual topology, and its cohomology groups $H^i(X(\mathbf{C}), \mathbf{Q})$ with rational coefficients can be defined by sheaf theory ; one also gets cohomology groups with compact support $H^i_c(X(\mathbf{C}), \mathbf{Q})$. There are natural maps (due to the fact that the usual topology is finer than the étale one) :

$$H^i(\overline{X}, \mathbf{Q}_\ell) \to H^i(X(\mathbf{C}), \mathbf{Q}) \otimes \mathbf{Q}_\ell \quad \text{and} \quad H^i_c(\overline{X}, \mathbf{Q}_\ell) \to H^i_c(X(\mathbf{C}), \mathbf{Q}) \otimes \mathbf{Q}_\ell.$$

The following basic result is the étale analogue of GAGA's theorems for coherent cohomology :

**Theorem 4.1.** (M.Artin) *The above maps are isomorphisms.*

The proof for $H^i$ can be found in [SGA 4, XVI.4, th. 4.1], in a more general setting ; the case of $H^i_c$ follows, cf. [SGA 4, XVII.5.3]. Note that, if $k = \mathbf{R}$, these isomorphisms are compatible with the action of $\Gamma_{\mathbf{R}}$ , i.e. with complex conjugation.

*Remark.* Here also, the proofs have to be made first for the finite sheaves $\mathbf{Z}/\ell^n\mathbf{Z}$ ; the case of $\mathbf{Z}_\ell$ (and then of $\mathbf{Q}_\ell$) follows.

## 4.3. Finite fields : Grothendieck's theorem

Suppose $k$ is finite, with $q = p^e$ elements, $p$ prime. In that case, there is a natural $k$-morphism $F : X \to X$, called the *Frobenius endomorphism* of $X$, defined as follows :

It is the identity on the topological space $X$, and its action on the structure sheaf $\mathcal{O}_X$ is $f \mapsto f^q$.

In particular, if $X$ is affine, and is defined by polynomial equations

$$f_\alpha(x_1, ..., x_n) = 0$$

with coefficients in $k$, then $F$ is the standard Frobenius map :

$$(x_1, ..., x_n) \mapsto (x_1^q, ..., x_n^q).$$

One of its main properties is that a $\overline{k}$-point $x$ of $X$ is $k$-rational if and only if it is fixed under $F$. Similarly, if $m$ is an integer $> 0$, and if $k_m$ denotes the subextension of $\overline{k}$ of degree $m$ over $k$, then $X(k_m)$ is the subset of $X(\overline{k})$ made up of the points fixed under the $m$-th iterate $F^m$ of $F$.

The morphism $F : X \to X$ is proper; hence it acts by functoriality on the cohomology spaces $H^i_c(\overline{X}, \mathbf{Q}_\ell)$, where $\ell$ is any prime number $\neq p$. Let us denote by $\mathrm{Tr}_i(F)$ the trace of this endomorphism, and define :

$$\mathrm{Tr}(F) = \sum_i (-1)^i \mathrm{Tr}_i(F).$$

This is the *Lefschetz number* of $F$, relative to the $\ell$-adic cohomology with proper support. *A priori*, it depends on the choice of $\ell$. In fact, it does not, because of the following result of Grothendieck ([Gr 64], see also [SGA $4\frac{1}{2}$, p.86, th.3.2]) :

**Theorem 4.2.** $\mathrm{Tr}(F) = |X(k)|$.

This also applies to the finite extensions of $k$. Hence :

**Corollary 4.3.** $\mathrm{Tr}(F^m) = |X(k_m)|$ *for every* $m \geqslant 1$.

*Remarks.*

1) Since $F : \overline{X} \to \overline{X}$ is a radicial morphism, it is an homeomorphism for the étale topology. Hence every eigenvalue of $F$ on $H^i_c(\overline{X}, \mathbf{Q}_\ell)$ is non-zero; for a more precise statement, see Theorem 4.5 below.

2) The theorem proved by Grothendieck, *loc.cit.*, is more general than Theorem 4.2 : it applies to every constructible $\mathbf{Q}_\ell$-sheaf, and gives $\mathrm{Tr}(F)$ as a sum of local traces at the points of $X(k)$.

3) Assume $k = \mathbf{F}_p$, to simplify notations. Then Corollary 4.3 is equivalent to saying that the Dirichlet series denoted by $\zeta_{X,p}(s)$ in §1.5 is equal

to $\prod_i \det(1 - p^{-s}F|H^i_c(\overline{X}, \mathbf{Q}_\ell))^{(-1)^{i+1}}$, which is a rational function of $p^{-s}$. Moreover, one has

$$N_X(p^e) = \sum_i (-1)^i \mathrm{Tr}_i(F^e)$$

for every $e \in \mathbf{Z}$ (and not merely for $e \geqslant 1$). In particular, $N_X(p^0)$ is equal to $\sum_i (-1)^i \dim H^i_c(\overline{X}, \mathbf{Q}_\ell)$, which is the *Euler-Poincaré characteristic* of $\overline{X}$.

## 4.4. The case of a finite field : the geometric and the arithmetic Frobenius

Keep the notation of §4.3. The Galois group $\Gamma_k = \mathrm{Gal}(\overline{k}/k)$ acts on each cohomology group $H^i_c(\overline{X}, \mathbf{Q}_\ell)$. In particular, the natural generator $\sigma = \sigma_q$ of $\Gamma_k$ acts by an automorphism (still denoted by $\sigma$), that is called the *arithmetic* Frobenius automorphism in order to distinguish it from the *geometric* Frobenius $F$ defined above. These two kind of "Frobenius automorphisms" are related by the following simple result (see [SGA 5, p.457], or [Ka 94, 24-25]) :

**Theorem 4.4.** *The arithmetic Frobenius and the geometric Frobenius are inverses of each other.*

In other words, we have $\sigma(F\xi) = F(\sigma\xi) = \xi$ for every $\xi \in H^i_c(\overline{X}, \mathbf{Q}_\ell)$.

A similar result holds for the cohomology groups $H^i(\overline{X}, \mathbf{Q}_\ell)$, with arbitrary support (and also for the cohomology with coefficients in $\mathbf{Z}/\ell^n\mathbf{Z}$).

*Example.* Suppose that $X$ is an abelian variety over $k$, and let $V_\ell(X)$ be its Tate $\mathbf{Q}_\ell$-module.

[Recall that $V_\ell(X) = \mathbf{Q} \otimes \varprojlim X[\ell^n]$, where $X[\ell^n]$ is the group of the $\ell^n$-division points of $X(\overline{k})$, i.e. the kernel of $\ell^n : X(\overline{k}) \to X(\overline{k})$ ; it is a $\mathbf{Q}_\ell$-vector space of dimension $2\dim X$.]

The Frobenius endomorphism $F : X \to X$ acts on $V_\ell(X)$ ; the arithmetic Frobenius $s$ also acts, and its action is the same as the action of $F$ (because $F$ and $s$ act in the same way on $X(\overline{k})$). The first cohomology group $H^1(\overline{X}, \mathbf{Q}_\ell)$ is the dual of $V_\ell(X)$ ; the action of $F$ on it is defined by functoriality, i.e. by transposition ; the action of $s$ is defined by *transport of structure*, i.e. by inverse transposition. This explains why the two actions are inverse of each other.[3]

[3]What this example suggests is that, if étale topology were expressed in terms of homology instead of cohomology, the two types of Frobenius would be the same.

## 4.5. The case of a finite field : Deligne's theorems

We keep the notation and hypotheses of §4.4 above.

Recall that a *$q$-Weil integer of weight* $w \in \mathbf{N}$ is an algebraic integer $\alpha$ such that $|\iota(\alpha)| = q^{w/2}$ for every embedding $\iota : \mathbf{Q}(\alpha) \to \mathbf{C}$. For instance a $q$-Weil integer of weight 0 is a root of unity (Kronecker).

*Remark.* In Deligne [De 80, §1.2.1], what we call a $q$-Weil integer of weight $w$ is called " an algebraic integer that is pure of weight $w$ relatively to $q$ ".

**Theorem 4.5.** (Deligne) *Let* $d = \dim X$.

a) *Every eigenvalue $\alpha$ of the geometric Frobenius $F$ acting on $H^i_c(\overline{X}, \mathbf{Q}_\ell)$ is a $q$-Weil integer of weight $\leqslant i$ ; if $i \geqslant d$, then $\alpha$ is divisible by $q^{i-d}$.*

b) *Assume that $X$ is proper and smooth. Then the characteristic polynomial of $F$ acting on $H^i_c(\overline{X}, \mathbf{Q}_\ell)$ has coefficients in $\mathbf{Z}$ and is independent of $\ell$ ; its roots are $q$-Weil integers of weight $i$.*

Assertion b) is the celebrated Weil conjecture. It is proved in [De 74] under the slightly restrictive assumption that $X$ is projective, instead of merely proper; the general case can be found in [De 80].

The proof of a) is given in [De 80, §3.3]; see also [Ka 94, pp.26-27]. The divisibility of $\alpha$ by $q^{i-d}$ is not explicitly stated in [De 80], but it follows from Cor. 3.3.8 which says that, if $i \geqslant d$, one has $v(\alpha) \geqslant (i-d)v(q)$ for every $p$-adic valuation $v$ of $\mathbf{Q}(\alpha)$.

Let us fix a prime $\ell \neq p$, and define the $i$-th Betti number $B_i$ of $X$ as the $\mathbf{Q}_\ell$-dimension of $H^i_c(\overline{X}, \mathbf{Q}_\ell)$ (recall that it is independent of $\ell$ in case b) above). Let $N_X(q)$ be, as usual, the number of elements of $X(k)$. By combining Theorem 4.2 and Theorem 4.5 we get :

**Theorem 4.6.** $N_X(q) = \sum_{i=0}^{i=2d} (-1)^i \nu_i$, *where* $\nu_i = \mathrm{Tr}(F | H^i_c(\overline{X}, \mathbf{Q}_\ell))$ *is an algebraic integer such that* $|\nu_i| \leqslant q^{i/2} B_i$.

(More precisely, all the Galois conjugates of $\nu_i$ have absolute value bounded by $q^{i/2} B_i$.)

Note that the "main term" in this formula is $\nu_{2d}$, where $d = \dim X$. Its value is easy to compute. Indeed, let us denote by $I_X$ the set of irreducible components of $\overline{X}$ of dimension $d$. It is proved in [SGA 4, XVIII.2.9] that $H^{2d}_c(\overline{X}, \mathbf{Q}_\ell)$ is canonically isomorphic[4] to $\mathbf{Q}_\ell(-d)^{I_X}$. Since $F$ acts by $q^d$ on $\mathbf{Q}_\ell(-d)$, this shows that we have $\nu_{2d} = eq^d$, where $e$ is the number of elements of $I_X$ that are $k$-rational, i.e. invariant under $\sigma_q$. By applying Theorem 4.5 we obtain the following estimate :

[4] Recall that $\mathbf{Q}_\ell(-d)$ is the $d$-th tensor power of $\mathbf{Q}_\ell(-1)$, and that $\mathbf{Q}_\ell(-1)$ is the $\mathbf{Q}_\ell$-dual of $\mathbf{Q}_\ell(1) = \mathbf{Q} \otimes \varprojlim \mu_{\ell^n}$, where $\mu_{\ell^n}$ is the group of $\ell^n$-th roots of unity in $\overline{k}$.

**Corollary 4.7.** *$|N_X(q) - eq^d| \leqslant (B - B_{2d})q^{d-\frac{1}{2}}$, where $B = \sum_i B_i$.*

This is especially useful when $X$ is geometrically irreducible, because $I_X$ has only one element, hence $e = 1$, and we get the bound

$$|N_X(q) - q^d| \leqslant (B-1)q^{d-\frac{1}{2}}.$$

*Remarks.*

1) When $X$ is given by equations of known degrees, one can give an explicit bound for $B$, cf. [Ka 01a].

2) When $X$ is proper and smooth, it follows from Theorem 4.5 (b) that, for every $i$, the number $\nu_i$ belongs to $\mathbf{Z}$ and is independent of the choice of $\ell$. The inequality $|\nu_i| \leqslant q^{i/2}B_i$, can then be improved to :

$$|\nu_i| \leqslant \frac{B_i}{2}[2q^{i/2}],$$

where [ ] is the floor function "integral part" ; moreover, the extreme cases $\nu_i = \frac{B_i}{2}[2q^{i/2}]$ and $\nu_i = -\frac{B_i}{2}[2q^{i/2}]$ can only occur when all the eigenvalues of $F$ on $H^i(\overline{X}, \mathbf{Q}_\ell)$ have real part $\frac{1}{2}[2q^{i/2}]$ or have real part $-\frac{1}{2}[2q^{i/2}]$. This follows from Theorem 4.10 below, applied to the characteristic polynomial of $F$ acting on $H^i_c(\overline{X}, \mathbf{Q}_\ell)$.

*Exercise.* Let $\alpha$ be a $q$-Weil integer of weight $i \geqslant 1$. Show that there exists an abelian variety $X$ over $k$ such that $\alpha$ is an eigenvalue of $F$ acting on $H^i(\overline{X}, \mathbf{Q}_\ell)$.

[Hint. By Honda-Tate's theorem, cf. [Ta 69], applied to an $i$-th root $\beta$ of $\alpha$, there exists an abelian variety $A$ over $k$ such that $\beta$ occurs as an eigenvalue of $F$ acting on $H^1(\overline{A}, \mathbf{Q}_\ell)$ ; choose for $X$ the product of $i$ copies of $A$.]

## 4.6. Improved Deligne-Weil bounds

As mentioned above, the bounds given by Deligne's theorem can be slightly improved by using the known relations between $q$-Weil numbers and totally real positive algebraic integers, cf. [Se 84, p.81]. Let us recall how this is done :

### 4.6.1. Totally real positive algebraic integers

Let $z$ be an element of a field of characteristic 0, for instance $\mathbf{C}$ or $\overline{\mathbf{Q}}_\ell$. Assume that $z$ is algebraic over $\mathbf{Q}$ and denote by $\mathrm{Tr}(z)$ its trace in the field extension $\mathbf{Q}(z)/\mathbf{Q}$ ; we have $\mathrm{Tr}(z) = \sum \iota(z)$, where $\iota$ runs through the embeddings $\mathbf{Q}(z) \to \mathbf{C}$.

One says that $z$ is *totally real* if $\iota(z)$ is real for every $\iota : \mathbf{Q}(z) \to \mathbf{C}$. If moreover we have $\iota(z) > 0$ for every $\iota$, $z$ is said to be *totally real positive*, that we shall abbreviate by *$z$ is totally positive*[5].

**Theorem 4.8.** (Siegel [Si 45]) *Let $z$ be a totally positive algebraic integer, and let $d(z) = [\mathbf{Q}(z) : \mathbf{Q}]$ be its degree. Then :*

1) $\mathrm{Tr}(z)/d(z) \geqslant 1$, *with equality only if* $z = 1$.

2) $\mathrm{Tr}(z)/d(z) \geqslant 3/2$ *if* $z \neq 1$, *with equality only if* $z = (3 \pm \sqrt{5})/2$.

*Proof.* Part 1) is easy : we have $\sum \iota(z) = \mathrm{Tr}(z)$ and $\prod \iota(z)$ is a strictly positive integer, hence is $\geqslant 1$. The standard inequality between arithmetic mean and geometric mean gives

$$\frac{1}{d(z)} \sum \iota(z) \geqslant \left(\prod \iota(z)\right)^{1/d(z)} \geqslant 1,$$

as desired. Moreover, one only has equality if all the $\iota(z)$ are equal to 1, i.e. if $z = 1$.

The proof of part 2) given in [Si 45] is less straightforward, but C. Smyth ([Sm 84]) has given a simpler and more general proof which also gives a complete (and finite!) list of the $z$'s such that $\mathrm{Tr}(z)/d(z) \leqslant 1.7719$; as a consequence, if $\mathrm{Tr}(z)/d(z) > 3/2$, then $\mathrm{Tr}(z)/d(z) \geqslant 5/3$, with equality only if $z$ is a conjugate of the cubic number $2+2\cos(2\pi/7)$, i.e. is such that $z^3 - 5z^2 + 6z - 1 = 0$. The constant 1.7719 has been improved several times, the 2008-record being 1.784109, cf. [AP 08]. Whether it can be brought up to $2 - \varepsilon$ for every $\varepsilon > 0$ is an open question that is discussed in [AP 08] under the name of the "Schur-Siegel-Smyth trace problem"; however, the natural bound for Smyth's method seems to be the number 1.898302..., rather than the number 2, as explained in [Se 98].

**Corollary 4.9.** *Let $A(x) = x^d - a_1 x^{d-1} + \cdots$ be a monic polynomial of degree $d > 0$, with coefficients in $\mathbf{Z}$. Assume that all the roots of $A$ are real $> 0$. Then :*

1 ) $a_1 \geqslant d$, *with equality if and only if* $A(x) = (x-1)^d$.

2) *If* $a_1 = d+1$, *then* $A(x)$ *is equal to one of the following two polynomials :*

$$(x-1)^{d-1}(x-2), \quad (x-1)^{d-2}(x^2 - 3x + 1).$$

[5] Note that here we follow the English tradition, where "positive" means $> 0$. Note also that "totally positive" has a different meaning in class field theory when one defines restricted ideal class groups : an element $z$ of a number field $K$ is called " totally positive in $K$ " if $\iota(z) > 0$ for every embedding $\iota$ of $K$ in $\mathbf{R}$. Note that this depends not only on $z$ but also on the field $K$ containing $z$; for instance, $-1$ is totally positive in $\mathbf{Q}(i)$.

*Proof.* If the statement is true for two polynomials, it is also true for their product. One may thus assume that $A$ is irreducible; if $z$ is one of its roots, then $z$ is a totally positive algebraic integer of degree $d$ and $\mathrm{Tr}(z) = a_1$; assertion 1) follows directly from part 1) of Theorem 4.8. If $a_1 = d+1$, part 2) of Theorem 4.8 shows that $(d+1)/d \geqslant 3/2$, hence $d = 1$ or 2. If $d = 1$, then $z = 2$ and $A(x) = x - 2$. If $d = 2$, then $\mathrm{Tr}(z)/d$ is equal to $3/2$, we have $z = (3 \pm \sqrt{5})/2$ and $A(x) = x^2 - 3x + 1$.

*Remark.* Using Smyth's results ([Sm 84, p.6]), one can also make a list of the $A$'s for which $a_1 = d+2$, or $a_1 = d+3$, ..., up to $a_1 = d+6$. For instance $a_1 = d+2$ is possible only when $A$ is one of the following seven polynomials:

$(x-1)^{d-1}(x-3)$, $(x-1)^{d-2}(x-2)^2$, $(x-1)^{d-3}(x-2)(x^2-3x+1)$, $(x-1)^{d-4}(x^2-3x+1)^2$, $(x-1)^{d-2}(x^2-4x+1)$, $(x-1)^{d-2}(x^2-4x+2)$, $(x-1)^{d-3}(x^3-5x^2+6x-1)$.

### 4.6.2. Relations between $q$-Weil integers and totally positive algebraic integers

Let $\alpha$ be a $q$-Weil integer of weight $w$, cf. §4.5; the element $\beta = q^w/\alpha$ of $\mathbf{Q}(\alpha)$ has the property that $\iota(\beta) = \overline{\iota(\alpha)}$ for every $\iota : \mathbf{Q}(\alpha) \to \mathbf{C}$; it is thus natural to denote it by $\overline{\alpha}$. Let $a = \alpha + \overline{\alpha} \in \mathbf{Q}(\alpha)$. It is clear that $a$ is totally real, and we may view $a/2$ as the "real part" of $\alpha$. We have

(4.6.2.1) $-2q^{w/2} \leqslant \iota(a) \leqslant 2q^{w/2}$ for every $\iota$.

Conversely, if $a$ is a totally real algebraic integer such that (4.6.2.1) is valid, then the two roots of the equation $x^2 - ax + q^w = 0$ are $q$-Weil integers of weight $w$. Note that these two roots are equal if and only if $a^2 = 4q^w$, in which case they are both equal to $q^{w/2}$ or both equal to $-q^{w/2}$.

Condition (4.6.2.1) can be stated in a different way. Let us write $2q^{w/2}$ as $2q^{w/2} = m + \varepsilon$, with $m \in \mathbf{N}$ and $0 \leqslant \varepsilon < 1$, i.e. $m = [2q^{w/2}]$ and $\varepsilon = \{2q^{w/2}\}$ with standard notation. It follows from (4.6.2.1) that:

(4.6.2.2) *The algebraic integers $z = m + 1 - a$ and $z' = m + 1 + a$ are totally positive.*

Moreover, if we denote by $\inf(z)$ (resp. $\inf(z')$) the smallest of all the real conjugates of $z$ (resp. of $z'$), we have

(4.6.2.3) $\inf(z) \geqslant 1 - \varepsilon$, *and* $\inf(z') \geqslant 1 - \varepsilon$.

Conversely, if $z, z'$ are totally positive algebraic integers with $z + z' = 2m + 2$, verifying conditions (4.6.2.2) and (4.6.2.3), and if we define $a$ by $a = m + 1 - z = z' - 1 - m$, the two roots of the equation $\alpha^2 - a\alpha + q^w = 0$ are $q$-Weil integers of weight $w$.

This correspondence allows us to transform statements on totally positive algebraic integers into statements on $q$-Weil integers. For instance :

**Theorem 4.10.** *Let $P(x) = x^n - b_1x^{n-1} + \cdots$ be a monic polynomial of degree $n > 0$, with coefficients in $\mathbf{Z}$. Assume that the roots of $P$ are $q$-Weil integers of weight $w$; put $m = [2q^{w/2}]$, as above. Then :*

a) $|b_1| \leqslant nm/2$;

b) *If $b_1 = nm/2$, then every root $\alpha$ of $P$ has a real part* (in the sense defined above) *equal to $m/2$, i.e. we have $\alpha^2 - m\alpha + q^w = 0$.*

c) *If $b_1 = nm/2 - 1$, there are only two possibilities :*

($c_1$) *All the roots of $P$ but two have real part $m/2$, the other two having real part $(m-1)/2$.*

($c_2$) *All the roots of $P$ but four have real part $m/2$, the other four having real part $m - z$ and $m - z'$ where $z, z' = (1 \pm \sqrt{5})/2$. This case is possible only when $\{2q^{w/2}\} \geqslant (\sqrt{5}-1)/2 = 0.61803...$*

(There are similar result when $b_1 = -nm/2$ or $b_1 = -nm/2 + 1$; they follow from b) and c) applied to the polynomial $(-1)^n P(-x)$.)

*Proof.* Assume first that neither $q^{w/2}$ nor $-q^{w/2}$ is a root of $P(x)$. In this case, the roots of $P$ come in pairs of conjugates; hence the degree $n$ is even, and we may write $P$ as $P(x) = \prod_{j=1}^{n/2}(x - \alpha_j)(x - \overline{\alpha}_j)$.

Define a polynomial $A(x) = x^{n/2} - a_1x^{n/2-1} + \cdots$ by the formula :

$$A(x) = \prod_{j=1}^{n/2}(x + \alpha_j + \overline{\alpha}_j - m - 1).$$

Its coefficients belong to $\mathbf{Z}$, and by (4.6.2.2) its roots are real $> 0$. By part 1) of Corollary 4.9, its first coefficient $a_1$ is $\geqslant d = n/2$. We have

$$a_1 = \sum_{j=1}^{n/2}(m+1-\alpha_j-\overline{\alpha}_j) = (m+1)n/2 - \sum(\alpha_j+\overline{\alpha}_j) = (m+1)n/2 - b_1.$$

We thus get $(m+1)n/2 - b_1 \geqslant d = n/2$, hence $b_1 \leqslant nm/2$. This proves a). The proof of b) is analogous.

In the general case, one writes $P(x)$ as

$$P(x) = (x - q^{w/2})^r(x + q^{w/2})^s P_1(x,)$$

with $r, s \in \mathbf{N}$ and $P_1(x) = \prod\ (x - \alpha_j)(x - \overline{\alpha}_j)$, as above. By applying to $P_1(x)$ what we have just proved, one gets :

$$a_1 - (r-s)q^{w/2} \leqslant (n - r - s)m/2.$$

Since $m \leqslant q^{w/2}$, this gives $a_1 \leqslant nm/2$, and one also sees that equality is possible only when $s = 0$ and all the $\alpha_j + \overline{\alpha}_j$ are equal to $m$. Note that, if $q^w$ is not a square, we have $r = s$ (since $q^{w/2}$ and $-q^{w/2}$ are Galois conjugates over $\mathbf{Q}$), hence $r = 0$ and all the real parts of the roots are equal to $m/2$; this last fact remains true if $q^w$ is a square because then $m = 2q^{w/2}$. This completes the proof of b).

Assertion c) follows from part 2) of Corollary 4.9 by a similar argument; note the restriction $\{2q^{w/2}\} \geqslant (\sqrt{5}-1)/2$ in case $c_2$), that comes from (4.6.2.3).

## 4.7. Examples

Let us review, from the point of view of étale cohomology, the examples given in Chap.1 and Chap.2. We keep the hypotheses of §4.3 and §4.4; in particular, the ground field $k$ is finite with $q$ elements and we put $\Gamma_k = \mathrm{Gal}(\overline{k}/k)$.

In order to simplify the notation, we write $H^i$ instead of $H^i_c(\overline{X}, \mathbf{Q}_\ell)$.

### 4.7.1. Examples of dimension 0

Assume $\dim X = 0$, and that $X$ is reduced. In that case, $X$ is finite étale over $k$, hence is determined by the $\Gamma_k$-set $\Omega = X(\overline{k})$. [If $X = \operatorname{Spec} k[x]/(f)$, as in §2.1, $\Omega$ is the set of roots of $f$ in $\overline{k}$.] There is a natural action of the Frobenius automorphism $\sigma = \sigma_q$ on $\Omega$; we have

$$N_X(q) = |\Omega^\sigma| = \text{ number of fixed points of } \sigma \text{ acting on } \Omega.$$

The cohomology groups $H^i$ are 0 for $i \neq 0$. The group $H^0$ is the vector space $C(\Omega, \mathbf{Q}_\ell)$ of $\mathbf{Q}_\ell$-valued functions on $\Omega$; the action of the geometric Frobenius on it is the map $\varphi(\omega) \mapsto \varphi(\sigma(\omega))$, while the arithmetic Frobenius acts by $\varphi(\omega) \mapsto \varphi(\sigma^{-1}(\omega))$, for every $\varphi \in C(\Omega, \mathbf{Q}_\ell)$. These two Frobenius automorphisms have the same eigenvalues, which are $q$-Weil numbers of weight 0, i.e. roots of unity. Their trace is equal to $|\Omega^\sigma|$ = number of fixed points of $\sigma$ acting on $\Omega$, cf. Theorem 4.2.

### 4.7.2. Examples of dimension 1

4.7.2.1. *The affine curve* $x^2 + y^2 = 0$, *cf.* §1.3. Here we assume that $p \neq 2$, so that $X$ is reduced. One finds that $\dim H^i = i$ if $0 \leqslant i \leqslant 2$ and $H^i = 0$ if $i > 2$. Moreover, the action of $F$ on $H^i$ is the following :

if $i = 1$ : the identity if $q \equiv 1 \pmod 4$, and minus the identity if not;

if $i = 2$ : $q\sigma$ , where $\sigma = 1$ if $q \equiv 1 \pmod 4$ and $\sigma$ has eigenvalues 1 and $-1$ if not.

The corresponding traces are 1 (or $-1$) and $2q$ (or 0). The Grothendieck-Lefschetz formula (Theorem 4.2) gives $N_X(q) = 2q - 1$ in the first case, and $N_X(q) = 1$ in the second case.

4.7.2.2. *Smooth projective curves.* Let $X$ be a smooth projective curve that is geometrically irreducible of genus $g$, and let $J$ be its Jacobian variety. The $H^i$ are 0 for $i > 2$, and for $i = 0, 1, 2$ they are as follows :

$H^0 = \mathbf{Q}_\ell$;

$H^1$ = dual of the Tate $\ell$-adic space $V_\ell(J)$ of $J$, cf. §4.4; its dimension is $2g$;

$H^2 = \mathbf{Q}_\ell(-1)$, dual of $\mathbf{Q}_\ell(1) = V_\ell(\mathbf{G}_m) = \mathbf{Q}_\ell \otimes \varprojlim \mu_{\ell^n}$.

The actions of $F$ on $H^0, H^1, H^2$ are the obvious ones : multiplication by $q$ (resp. by 1) on $H^2$ (resp. on $H^0$), and action on $H^1$ by the transpose of the Frobenius endomorphism $F_J$ of $J$. The Grothendieck-Lefschetz formula (due to Weil in this case, cf. [We 48 a,b]) gives :

$$N_X(q) = q + 1 - a,$$

where $a = \mathrm{Tr}(F_J | V_\ell(J))$ is the trace of the endomorphism $F_J$ of $J$. As Weil himself showed, $a$ is the sum of $2g$ $q$-Weil integers of weight 1 (that explains the terminology). This shows that $|a| \leqslant 2gq^{\frac{1}{2}}$, hence :

$$|N_X(q) - q - 1| \leqslant 2gq^{\frac{1}{2}}.$$

Using §4.6.2, we also get the improved bound :

$$|N_X(q) - q - 1| \leqslant gm, \quad \text{with } m = [2q^{\frac{1}{2}}],$$

the equality $N_X(q) - q - 1 = gm$ being only possible if every eigenvalue $\alpha$ of $F$ on $H^1$ is such that $\alpha^2 + m\alpha + q = 0$; when $m$ is prime to $q$, this implies that $J$ is isogenous to a product of $g$ copies of an elliptic curve with $q + 1 + m$ rational points.

The above bounds seem reasonably sharp when $q$ is large with respect to $g$, for instance for fixed $g$ and $q \to \infty$; this is at least what happens for $g = 1$ and 2, and there are encouraging partial results for $g = 3$, cf. [La 02]. On the other hand, when $g$ is large with respect to $q$, there are much better bounds, due to Ihara, Drinfeld, Vlăduts and others; see e.g. the references and the numerical tables of [GV 09].

### 4.7.3. Examples of dimension 2

Suppose $X$ is a projective smooth geometrically irreducible $k$-surface. We then have $H^0 = \mathbf{Q}_\ell$ and $H^4 = \mathbf{Q}_\ell(-2)$, with standard notation (recall that $\mathbf{Q}_\ell(-2)$ is the second tensor power of $\mathbf{Q}_\ell(-1)$, where $\mathbf{Q}_\ell(-1)$ denotes the dual of the Tate space $V_\ell(\mathbf{G}_m)$). The action of $F$ on these spaces is by multiplication by 1 and by $q^2$ respectively. Hence $H^0$ and $H^4$ contribute $1+q^2$ to $N_X(q)$. Moreover, if $NS$ denotes the Néron-Severi group of $\overline{X}$, there is a natural embedding of $NS \otimes \mathbf{Q}_\ell(-1)$ into $H^2$, given by the "cycle map", see [SGA $4\frac{1}{2}$, pp.129-153]. The action of $F$ on that space is $\sigma_q^{-1} \otimes q$, where $\sigma_q$ is the action of the arithmetic Frobenius on $NS$; its trace is thus equal to $q\mathrm{Tr}(\sigma_q^{-1})$ which is the same as $q\mathrm{Tr}(\sigma_q)$ since a permutation matrix has the same trace as its inverse. In the cases considered in §2.3.2 and §2.3.3, the surface $X$ is geometrically rational, which implies $NS \otimes \mathbf{Q}_\ell(-1) = H^2$, and $H^1 = H^3 = 0$. We thus get

$$N_X(q) = q^2 + q\mathrm{Tr}(\sigma_q) + 1,$$

as stated in §2.3.3.

## 4.8. Variation with $p$

### 4.8.1. Notation

Let us go back to the setting of §3.1, where $K$ is a number field, $O_K$ is the ring of integers of $K$ and $V_K$ is the set of non-archimedean places of $K$.

Let $X$ be a separated scheme of finite type over $O_K$, and denote by $X_K$ the corresponding $K$-variety, i.e. the generic fiber of $X \to \operatorname{Spec} O_K$. For every $v \in V_K$, we have an injection $\operatorname{Spec} \kappa(v) \to \operatorname{Spec} O_K$; hence, by pull-back, a $\kappa(v)$-scheme $X_v$, which is sometimes called the "reduction mod $\mathfrak{p}_v$" of $X$ :

$$\begin{array}{ccccc} X_v & \to & X & \leftarrow & X_K \\ \downarrow & & \downarrow & & \downarrow \\ \operatorname{Spec} \kappa(v) & \to & \operatorname{Spec} O_K & \leftarrow & \operatorname{Spec} K. \end{array}$$

When $K = \mathbf{Q}$, $V_K$ is the set of prime numbers, and $X_v$ is the $\mathbf{F}_p$-scheme denoted by $X_p$ in §1.2.

We want to compare the étale $\ell$-adic cohomology groups of $\overline{X_K}$ and $\overline{X_v}$ (with $\ell \neq p_v$). In order to have a short enough notation, let us call " $A$ " any one of the $\ell$-adic groups $\mathbf{Z}/\ell^n\mathbf{Z}$, $\mathbf{Z}_\ell$ and $\mathbf{Q}_\ell$, so that $H^i(\overline{X_K}, A)$ and $H^i_c(\overline{X_K}, A)$ make sense, and let us use a similar notation for $X_v$.

### 4.8.2. Comparison theorems for cohomology with proper support

The first comparison theorem is :

**Theorem 4.11.** *For every $\ell$, there exists a finite subset $S_\ell$ of $V_K$, containing the places $v$ with $p_v = \ell$, such that, for every $i$ and for every $v \notin S_\ell$, the group $H^i_c(\overline{X_v}, A)$ is isomorphic to $H^i_c(\overline{X_K}, A)$.*

This fits with the intuitive idea that most fibers of a reasonable map (such as $X \to \operatorname{Spec} O_K$) should have similar topological properties, hence isomorphic cohomology. However the conclusion "is isomorphic to" is not precise enough to be useful; indeed, when $A = \mathbf{Q}_\ell$, it merely says that the $\mathbf{Q}_\ell$-vector spaces $H^i_c(\overline{X_v}, \mathbf{Q}_\ell)$ and $H^i_c(\overline{X_K}, \mathbf{Q}_\ell)$ have the same dimension. We need more, namely a *canonical* isomorphism, compatible with the local Galois action. To do so, we have to be more precise about the choice of an algebraic closure of $\kappa(v)$, since the very definition of $\overline{X_v}$ depends on that choice.

This is done as follows : For every $v \in V_K$, let us choose an extension $\overline{v}$ of $v$ to $\overline{K}$. The residue field $\kappa(\overline{v})$ of $\overline{v}$ is an algebraic closure of $\kappa(v)$; this is the algebraic closure we choose for defining $\overline{X_v}$. One can then define a canonical map (called the *specialization map*) :

$$r^i_{\overline{v}} \ :\ H^i_c(\overline{X_v}, A) \ \to\ H^i_c(\overline{X_K}, A),$$

and the refined form of Theorem 4.11 is :

**Theorem 4.12.** *For every $\ell$, there exists a finite subset $S_\ell$ of $V_K$, containing the places $v$ with $p_v = \ell$, such that, for every $i$ and for every $v \notin S_\ell$, the map $r^i_{\overline{v}}$ is an isomorphism.*

As in §3.2, let $D_{\overline{v}}$ be the *decomposition subgroup* of $\overline{v}$, i.e. the subgroup of $\Gamma_K = \operatorname{Gal}(\overline{K}/K)$ made up of the elements $\gamma \in \Gamma_K$ such that $\gamma(\overline{v}) = \overline{v}$; this group acts on $\kappa(\overline{v})$, and we have an exact sequence

$$1 \ \to\ I_{\overline{v}} \ \to\ D_{\overline{v}} \ \to\ \Gamma_{\kappa(v)} \ \to\ 1,$$

where $I_{\overline{v}}$ is the *inertia subgroup* of $\overline{v}$, and $\Gamma_{\kappa(v)} = \operatorname{Gal}(\kappa(\overline{v})/\kappa(v))$; let $\sigma_v$ be the canonical generator (i.e. the Frobenius element) of $\Gamma_{\kappa(v)}$.

The group $D_{\overline{v}}$ acts on $H^i_c(\overline{X_v}, A)$ via its quotient $\Gamma_{\kappa(v)}$; since it is a subgroup of $\Gamma_K$, it also acts on $H^i_c(\overline{X_K}, A)$. The specialization map $r^i_{\overline{v}}$, being canonical, commutes with these actions.

**Theorem 4.13.** *Let $S_\ell$ be as in Theorem* 4.12, *and assume $v \notin S_\ell$. Then :*

i) *The action of $\Gamma_K$ on $H^i_c(\overline{X_K}, A)$ is unramified at $v$* (i.e. $I_{\overline{v}}$ acts trivially).

ii) *If we identify* $H^i_c(\overline{X_K}, A)$ *with* $H^i_c(\overline{X_v}, A)$ *via* $r^i_{\overline{v}}$, *the action of* $\sigma_v$ *on* $H^i_c(\overline{X_K}, A)$ *coincides with the action of the arithmetic Frobenius* (cf. §4.4) *on* $H^i_c(\overline{X_v}, A)$.

iii) $|X(\kappa(v))| = \sum_i (-1)^i \mathrm{Tr}(\sigma_{\overline{v}}^{-1} | H^i_c(\overline{X_K}, \mathbf{Q}_\ell))$.

Assertions i) and ii) follow from the fact that $r^i_{\overline{v}}$ commutes with the action of $D_{\overline{v}}$. Assertion ii) follows from ii), combined with Theorems 4.2 and 4.3.

*Remark.* Theorems 4.11, 4.12 and 4.13 i), ii) are also valid for cohomology with arbitrary support; we shall not need them.

### 4.8.3. References

The results stated in §4.8.2 are merely a watered-down version of those the reader will find in [SGA 4], [SGA $4\frac{1}{2}$] and [SGA 5]. The basic steps of their proofs are :

a) Construction of the direct image sheaf $R^i f_! A$, where $f$ denotes the projection map $X \to \mathrm{Spec}\, O_K$ ; proof that this construction commutes with base change ([SGA $4\frac{1}{2}$, p.49, th.5.4]).

b) Proof that $R^i f_! A$ is "constructible" ([SGA $4\frac{1}{2}$, p.50, th.6.2]), hence is lisse on a dense open set of $\mathrm{Spec}\, O_K$ (namely the complement of the set $S_\ell$ above).

### 4.8.4. The exceptional set $S_\ell$

In some applications, one needs more information on the set $S_\ell$ occurring in theorem 4.12, and especially on the way it varies with $\ell$. By definition, $S_\ell$ contains the set $V_K(\ell)$ of $v$'s with $p_v = \ell$. A theorem of Katz and Laumon ([KL 86, th.3.1.2]) implies that one can take $S_\ell = \Sigma \cup V_K(\ell)$, where $\Sigma$ is finite and independent of $\ell$. In the special case where the $K$-variety $X_K$ is proper and smooth, one can choose for $\Sigma$ the set of $v$'s over which $X \to \mathrm{Spec}\, O_K$ is not proper and smooth, cf. [SGA $4\frac{1}{2}$, p.62, th.3.1] and [Mi 80, p.230, Cor.3.2].

CHAPTER 5

# AUXILIARY RESULTS ON GROUP REPRESENTATIONS

This chapter contains several results on the linear representations of a group $G$; we shall apply them in Chapter 6 to $G = \mathrm{Gal}(\overline{\mathbf{Q}}/\mathbf{Q})$.

## 5.1. Characters with few values

### 5.1.1. Grothendieck groups and characters

Let us first recall some definitions and notation (cf. [A VIII, §§11,20,21]).

5.1.1.1. *G-modules.* In what follows, $G$ is a group, and $K$ is a field of characteristic 0 (except in 5.1.1.4 below).

Let $K[G]$ be the *group algebra* of $G$ over $K$; its elements are the formal sums $\sum_{g \in G} \lambda_g g$, where the coefficients $\lambda_g$ belong to $K$ and are almost all equal to 0. We shall be interested in the category $\mathcal{C}_{G,K}$ of the $K[G]$-modules of finite dimension over $K$. An object $V$ of that category is a finite-dimensional $K$-vector space endowed with a homomorphism $G \to \mathrm{Aut}(V)$; by choosing a basis $\{e_1, ..., e_n\}$ of $V$, one may view it as a homomorphism $\rho : G \to \mathbf{GL}_n(K)$, i.e. as a $K$*-linear representation* of $G$. The morphisms are the linear maps that commute with the action of $G$.

Every $V$ has a composition series

$$0 = V_0 \subset V_1 \subset \cdots \subset V_m = V,$$

where the successive quotients $S_i = V_i/V_{i-1}$ are simple, i.e. give irreducible representations of $G$. The direct sum of the $S_i$ is independent (up to isomorphism) of the chosen composition series; it is a semisimple module, which is called *a semisimplification* [1] of $V$, and is denoted by $V^{ss}$.

5.1.1.2. *The representation ring.* Let $R_K(G)$ be the Grothendieck group associated with the category $\mathcal{C}_{G,K}$; if $V$ is an object of $\mathcal{C}_{G,K}$, its image in $R_K(G)$ is denoted by $[V]$. The set of all $[V]$ is an additive submonoid of $R_K(G)$, that we shall denote by $R_K(G)^+$. Every element of $R_K(G)$ can be written as a difference $[V_1] - [V_2]$. If

$$0 \to V' \to V \to V'' \to 0$$

[1] One sometimes calls it *the* semisimplification of $V$, by the same abuse of language as in "the" algebraic closure of a field, or "the" universal covering of a topological space.

is an exact sequence in $\mathcal{C}_{G,K}$, then $[V] = [V'] + [V'']$. As a consequence, we have $[V] = [V^{ss}]$ for every $V$, and $[V] = [V'] \iff V^{ss} \simeq V'^{ss}$. If $S_\alpha$ is a system of representatives of the simple objects of $\mathcal{C}_{G,K}$, the elements $[S_\alpha]$ make up a $\mathbf{Z}$-basis of $R_K(G)$. There is a structure of commutative ring on $R_K(G)$; its multiplication law is given by the rule $[V].[V'] = [V \otimes V']$. Its unit element corresponds to the module $V = K$, with trivial action of $G$; there is a natural homomorphism $\deg : R_K(G) \to \mathbf{Z}$, characterized by $[V] \mapsto \dim_K V$. This ring is called the *representation ring* of $G$ over $K$. It has more structure than being merely a commutative ring :

a) It has an *involution* $x \mapsto x^*$, given by the equation $[V]^* = [V^*]$, where $V^*$ is the $K$-dual of $V$, with its natural action of $G$.

b) It has $\lambda$-operations $x \mapsto \lambda^k x$, and also $\Psi^k$ operations $(k \geqslant 0)$, cf. [SGA 6, exposé 5], and [Se 78, §9.1, exerc. 3].

One can combine a) and b) to define $\Psi^{-k}$, for $k \geqslant 0$, by the formula $\Psi^{-k}(x) = \Psi^k(x^*) = \Psi^k(x)^*$; in particular $\Psi^{-1}(x) = x^*$.
These operations enjoy various properties, for instance :

$b_1$) $\lambda^k[V] = [\wedge^k V]$ for $k \geqslant 0$;

$b_2$) The $\Psi^k$ are ring-endomorphisms, and $\Psi^k \circ \Psi^{k'} = \Psi^{k+k'}$.

$b_3$) $\Psi^0 x = \deg(x).1$ and $\lambda^0 x = 1$.

$b_4$) $\Psi^1 x = \lambda^1 x = x$.

$b_5$) $\Psi^2 x + 2\lambda^2 x = x^2$.

5.1.1.3. *Characters.* If $V$ is as above, the function $\mathrm{Tr}_V : G \to K$ defined by $\mathrm{Tr}_V(g) = \mathrm{Tr}(g|V)$ is the *character* of $V$. It is a class function on $G$ : it takes the same value on conjugate elements. If we denote the vector space of all such functions by $\mathrm{Cl}(G, K)$, the map $[V] \mapsto \mathrm{Tr}_V$ extends by $K$-linearity to a map

$$\theta : K \otimes_{\mathbf{Z}} R_K(G) \to \mathrm{Cl}(G, K).$$

It is well known that $\theta$ is injective, cf. e.g. [A VIII, §20, part a) of cor. to prop.6]. In particular, if $V$ and $V'$ are such that $\mathrm{Tr}_V = \mathrm{Tr}_{V'}$, then $V^{ss}$ is isomorphic to $V'^{ss}$ : semisimple representations are detected by their character.

An element of $\mathrm{Cl}(G, K)$ that belongs to $\theta(R_K(G))$ is called a *virtual character* (or sometimes a *generalized character*) of $G$ over $K$.

If $x \in R_K(G)$ has character $f : G \to K$, and if $k$ is an integer, there is a simple formula for the character $\Psi^k f$ of $\Psi^k x$, namely :

$$\Psi^k f(g) = f(g^k).$$

5.1.1.4. *Characteristic p.* If $K$ were of characteristic $p > 0$, the character defined above would not be sufficient to detect equality in $R_K(G)$, since a direct sum of $p$ copies of the same $V$ has a character that is 0. However, there is a simple way out, due to Brauer. Assume for simplicity that $K$ is algebraically closed, and choose a ring $A$ with a surjective morphism $s : A \to K$ together with a homomorphism $r : K^\times \to A^\times$ such that $s \circ r$ is the identity on $K^\times$ ("multiplicative representatives") ; assume further that the sequence $0 \to A \xrightarrow{p} A \xrightarrow{s} K \to 0$ is exact and that $\bigcap_{n \geqslant 1} p^n A = 0$.

[Example of such an $A$ : the ring of infinite Witt vectors with coefficients in $K$, cf. [Se 62, Chap.II, §6]. The map $s : A \to K$ is $(x_0, x_1, ...) \mapsto x_0$ and the map $r : K^\times \to A^\times$ is $x \mapsto (x, 0, 0, ...)$.]

Let now $V$ be as above, and let $g$ be an element of $G$; let $\alpha_1, ..., \alpha_n$ be the eigenvalues of $g$ acting on $V$, where $n = \dim V$. Define the *Brauer character* $\mathrm{Tr}_V^{\mathrm{Br}}(g)$ of $g$ by the formula :

$$\mathrm{Tr}_V^{\mathrm{Br}}(g) = \sum_{i=1}^{i=n} r(\alpha_i).$$

The function $\mathrm{Tr}_V^{\mathrm{Br}} : G \to A$ is a class function on $G$; by composition with the projection $s : A \to K$ it gives the standard trace $\mathrm{Tr}_V$.

Let us denote by $\mathrm{Cl}(G, A)$ the $A$-module of all class functions on $G$ with values in $A$. It is not difficult to prove that the map $A \otimes_{\mathbf{Z}} R_K(G) \to \mathrm{Cl}(G, A)$ defined by $a \otimes [V] \mapsto a\,\mathrm{Tr}_V^{\mathrm{Br}}$ is injective (use [A VIII, §20, part b) of cor. to prop.6]). In particular, the map

$$\mathrm{Tr}^{\mathrm{Br}} : R_K(G) \to \mathrm{Cl}(G, A)$$

is injective; in other words, *two semisimple representations of G that have the same Brauer character are isomorphic.*

*Exercise.* (Brauer-Nesbitt) Let $V$ and $V'$ be such that the characters of $\wedge^k V$ and $\wedge^k V'$ are equal for every $k \in \mathbf{N}$. Show that $V^{ss} \simeq V'^{ss}$. [This is true in any characteristic.]

### 5.1.2. Characters of quotient groups : statements

**Theorem 5.1.** *Let $x$ be an element of $R_K(G)$ and let $f = \theta(x) \in \mathrm{Cl}(G, K)$ be its character. Assume that $f$ takes only finitely many values. Then there exists a normal subgroup $N$ of $G$, with $(G : N) < \infty$, such that $x$ belongs to the image of the natural map $R_K(G/N) \to R_K(G)$. In particular, $f$ is constant on every $N$-coset of $G$.*

The proof of a somewhat stronger result will be given in §5.1.3 and §5.1.4.

*Remark.* If $x = [V]$ is *effective*, i.e. belongs to $R_K(G)^+$, the same is true for the element of $R_K(G/N)$ it comes from : this will follow from the proof. In case $V$ is semisimple, this is equivalent to saying that the normal subgroup $N$ acts trivially on $V$.

A similar result was already proved by Weil in 1934 (cf. [We 34]) ; as he explains in his *Commentaires,* he had in mind a possible application to what was later called "Tannaka theory" for linear representations of the fundamental group, and for vector bundles.[2]

**Theorem 5.2.** *Let $f$ be as in Theorem 5.1. If the values of $f$ are contained in $\{-1, 0, 1\}$, there are only three possibilities :*

*a)* $f = 0$;

*b)* *$f$ is a homomorphism of $G$ into $\{\pm 1\}$;*

*c)* *$-f$ is a homomorphism of $G$ into $\{\pm 1\}$.*

*Proof.* Thanks to Theorem 5.1, we may assume that $G$ is finite. By enlarging $K$, we may also assume that $K$ is algebraically closed. One then writes $f$ as an integral linear combination $\Sigma\, n_i \chi_i$ of distinct irreducible characters $\chi_i$. The orthogonality formulae of characters show that the mean value of $f^2$ on $G$ is $\Sigma\, n_i^2$ ; since $f^2$ takes values in $\{0, 1\}$, its mean value belongs to $[0, 1]$ ; since it is an integer, this shows that either $f = 0$ or $f^2 = 1$. If the second case occurs, the sum of the $n_i^2$ is 1, which means that $f$ or $-f$ is equal to one of the $\chi_i$, i.e. either $f$ or $-f$ is an irreducible character. Since the value of that character at 1 is 1, either $f$ or $-f$ is a homomorphism of $G$ into $K^\times$, with values in $\{\pm 1\}$.

**Corollary 5.3.** *Let $f$ be as in Theorem 5.1, and assume that every value of $f$ is* 0 *or* 1*. Then either $f = 0$ or $f = 1$.*

This follows from Theorem 5.2, since case *c)* is excluded, and case *b)* is possible only when $f = 1$.

This corollary can be restated as :

**Corollary 5.4.** *The only idempotents of the representation ring $R_K(G)$ are* 0 *and* 1. (Equivalently : $\operatorname{Spec} R_K(G)$ *is connected.*)

Indeed, an idempotent of $R_K(G)$ corresponds, via $\theta$, to an $f$ with $f^2 = f$, i.e. to an $f$ with values in $\{0, 1\}$.

---

[2]His paper may well be the first one that introduces the representation ring of an infinite discrete group.

*Remark.* One may ask whether Theorem 5.2 can be extended to the case where $f$ takes values in a slightly different set, such as $\{0,3\}$. This can indeed be done, but the result is less simple, cf. §5.1.5 below.

*Exercises.*

1) Let $\mathfrak{g}$ be a Lie algebra over $K$ and let $R_K[\mathfrak{g}]$ be its representation ring. Show that $\operatorname{Spec} R_K[\mathfrak{g}]$ is connected.
[Hint. Reduce the question to the case where $\mathfrak{g}$ is finite dimensional and $K = \mathbf{C}$. Use the fact (cf. [LIE III, §6, th.3]) that there exists a simply connected complex Lie group $G$ such that $\operatorname{Lie} G \simeq \mathfrak{g}$ and observe that $R_K[\mathfrak{g}]$ embeds in $R_K[G]$.]

2) Assume $K$ has characteristic $p > 0$. Show that Corollary 5.4 remains valid, i.e. that $\operatorname{Spec} R_K(G)$ is connected.
[Hint. Reduce to the case where $K$ is algebraically closed, so that an idempotent element $x$ of $R_K(G)$ has a Brauer character $f$ (see end of §5.1.1) with values in $\{0,1\}$. Show that $f$ is constant, i.e. that $f(g) = f(1)$ for every $g \in G$. To do so, one may assume that $G$ is generated by $g$, hence is cyclic, in which case $x$ can be lifted to characteristic 0 and one applies Corollary 5.3 to that lift.]

### 5.1.3. First part of the proof of Theorem 5.1

**Theorem 5.5.** *Let $x$ be an element of $K \otimes_{\mathbf{Z}} R_K(G)$ and let $f \in \operatorname{Cl}(G,K)$ be its character. Assume that $f$ takes only finitely many values. Then there exists a normal subgroup $N$ of $G$, with $(G:N) < \infty$, such that $f$ is constant on every $N$-coset of $G$.*

Note that the condition on $x$ is less restrictive than in Theorem 5.1, where we asked that $x$ belongs to $R_K(G)$. On the other hand, we are not claiming yet that $x$ comes from $K \otimes_{\mathbf{Z}} R_K(G/N)$ ; this will only be proved in the next section.

*Proof.* We may assume that $K$ is algebraically closed. Let us write $f$ as $f = \Sigma a_i \chi_i$, with $a_i \in K$, the $\chi_i$ being the characters of some linear representations $\rho_i : G \to \mathbf{GL}_{n_i}(K)$. Let $\rho = (\rho_i) : G \to \prod \mathbf{GL}_{n_i}(K)$ be the homomorphism defined by the $\rho_i$'s. We may assume that $\rho$ has trivial kernel, so that $G$ can be identified with a subgroup of $\prod \mathbf{GL}_{n_i}(K)$. The function $f$ is the restriction to $G$ of the function

$$F : \prod \mathbf{GL}_{n_i}(K) \to K \quad \text{defined by} \quad F(g_i) = \sum a_i \operatorname{Tr}(g_i).$$

Let $\overline{G}$ be the Zariski closure of $G$ in $\prod \mathbf{GL}_{n_i}$. It is an algebraic subgroup of $\prod \mathbf{GL}_{n_i}$. Let $\overline{G}^0$ be the identity component[3] of $\overline{G}$ and let $S$ be the set of values of $f$. Since the polynomial function $F$ maps $G$ into the finite set $S$, it

[3] i.e. the connected component of the identity.

also maps $\overline{G}$ into $S$ and hence it is constant on every connected component of $\overline{G}$, i.e. it is constant on every $\overline{G}^0$-coset of $\overline{G}$. We may thus choose for $N$ the intersection $G \cap \overline{G}^0$ ; it is a subgroup of $G$ whose index is equal to the number of connected components of $\overline{G}$.

### 5.1.4. Characters of quotient groups

Let $N$ be a normal subgroup of $G$ (not necessarily of finite index). There is a natural embedding of the category $\mathcal{C}_{G/N,K}$ into the category $\mathcal{C}_{G,K}$ : every $G/N$-module can be viewed as a $G$-module on which $N$ acts trivially. This embedding gives rise to an injective ring homomorphism $R_K(G/N) \to R_K(G)$. We shall use it to identify $R_K(G/N)$ with a subring of $R_K(G)$; same convention for $K \otimes R_K(G/N) \to K \otimes R_K(G)$.

There is a natural additive projection $s : R_K(G) \to R_K(G/N)$. It is defined as follows :

Let $V$ be a semisimple object of $\mathcal{C}_{G,K}$ ; since $N$ is contained in $G$, one may view $V$ as an $N$-module, and since $N$ is normal, the $N$-module so obtained is semisimple (see e.g. [Se 94, Lemme 5]). Let $V^N$ be the subspace of $V$ fixed by $N$ ; the group $G/N$ acts on $V^N$, and we thus get a semisimple $G/N$-module. The map $s$ alluded to above is characterized by the formula

$$s([V]) = [V^N],$$

valid for every semisimple $V$. By linearity, we also have a map (still denoted by $s$) : $K \otimes R_K(G) \to K \otimes R_K(G/N)$ ; it is clear that $s(x) = x$ for every $x \in R_K(G/N)$.

If $x$ is an element of $K \otimes R_K(G)$, and $f \in \mathrm{Cl}(G, K)$ is its character, we shall denote by $x^N$ and $f^N$ the corresponding elements of $K \otimes R_K(G/N)$ and $\mathrm{Cl}(G/N, K)$.

*Remark.* When $N$ is finite, there is an explicit formula for $f^N$, namely

$$f^N(\gamma) = \frac{1}{|N|} \sum_{g \mapsto \gamma} f(g),$$

i.e. $f^N$ is the average of the $N$-translates[4] of $f$.
[Hint. Use the fact that, for every $V$, the element $\frac{1}{|N|} \sum_{n \in N} n$ of $K[N]$ defines a $G$-invariant projection of $V$ onto $V^N$.]

---

[4]Similarly, if $K = \mathbf{C}$, if $G$ is a topological group, $N$ a compact subgroup of $G$, and $R_K(G)$ is replaced by its topological analogue (relative to continuous representations), then $f^N$ is given by the same formula as above, except that the mean value is replaced by an integral over an $N$-coset.

**Theorem 5.6.** *Let $x$ be an element of $K \otimes R_K(G)$ and let $f \in \mathrm{Cl}(G, K)$ be its character. Assume that $f$ is constant on every $N$-coset of $G$. Then $f^N = f$, i.e. $x$ belongs to the subring $K \otimes R_K(G/N)$ of $K \otimes R_K(G)$.*

**Corollary 5.7.** *Assume $x$ and $f$ have the properties of Theorem* 5.6. *If $x$ belongs to $R_K(G)$* (resp. to $R_K(G)^+$), *then $x$ belongs to $R_K(G/N)$* (resp. to $R_K(G/N)^+$).

This follows from the easily checked equalities :

$$R_K(G) \cap K \otimes R_K(G/N) = R_K(G/N)$$

and

$$R_K(G)^+ \cap R_K(G/N) = R_K(G/N)^+.$$

*Proof of Theorem* 5.6.

Let us write $x$ as a finite sum $x = \sum_{i \in I} \lambda_i [V_i]$ with $\lambda_i \in K^\times$, the $V_i$ being simple, and pairwise non-isomorphic . By assumption, we have

$$\sum_{i \in I} \lambda_i \mathrm{Tr}_{V_i}((n-1)g) = 0$$

for every $g \in G$ and every $n \in N$ ; by linearity, this shows that

$$\sum_{i \in I} \lambda_i \mathrm{Tr}_{V_i}((n-1)\alpha) = 0$$

for every $n \in N$ and every $\alpha \in K[G]$. Let $i$ be an element of $I$. By the Jacobson density theorem ([A VIII, §5.5]), there exists $\alpha \in K[G]$ whose image in $\mathrm{End}(V_i)$ is 1, and whose image in $\mathrm{End}(V_{i'})$ is 0 if $i' \neq i$. If we apply the formula above to such an $\alpha$, we get $\lambda_i \mathrm{Tr}_{V_i}(n-1) = 0$, i.e. $\mathrm{Tr}_{V_i}(n-1) = 0$. This shows that the character of the $N$-module $V_i$ is constant ; since $V_i$ is $N$-semisimple, this implies that $N$ acts trivially on $V_i$, hence $[V_i] = [V_i^N]$ for every $i \in I$, and we have $x = x^N$, as claimed.

*End of the proof of Theorem* 5.1.

By Theorem 5.5, there exists a normal subgroup $N$ of $G$, of finite index, such that the character $f$ of $x$ is invariant by $N$-translations. By Theorem 5.6 and Corollary 5.7, this implies that $x = x^N$, i.e. that $x$ belongs to $R_K[G/N]$.

*Exercise (generalization of Theorem 5.6).* Let $A$ be a $K$-algebra and denote by $R_K(A)$ the Grothendieck group of the $A$-modules that are finite dimensional over $K$. The character of such a module is a linear form on $A$; this gives a homomorphism $\theta : K \otimes_{\mathbf{Z}} R_K(A) \rightarrow A'$, where $A'$ is the dual of the vector space $A$.

i) Show that $\theta$ is injective.

ii) Let $\mathfrak{n}$ be a two-sided ideal of $A$ and let $B = A/\mathfrak{n}$. There is a natural injection of $K \otimes R_K(B)$ into $K \otimes R_K(A)$. Show that an element of $K \otimes R_K(A)$ belongs to $K \otimes R_K(B)$ if and only if its character vanishes on $\mathfrak{n}$.

[The special case where $A = K[G]$ and $B = K[G/N]$ is Theorem 5.6. The proof in the general case is the same.]

### 5.1.5. Complement to Theorem 5.2 : virtual characters with set of values $\{0, p\}$.

We have seen (cf. Theorem 5.2) that there are very few virtual characters of a finite group $G$ whose set of values is $\{-1, 1\}$, namely the surjective homomorphisms $f : G \to \{\pm 1\}$ and their negatives. By replacing $f$ by $f + 1$, this also gives a list of the virtual characters with set of values $\{0, 2\}$ ; if one normalizes them by asking that $f(1) = 0$, they correspond to the subgroups $N$ of index 2 of $G$, via the rule :

$$f(x) = 0 \text{ if } x \in N \quad \text{and} \quad f(x) = 2 \text{ if } x \notin N.$$

It is natural to ask what happens when $\{0, 2\}$ is replaced by $\{0, p\}$, where $p$ is an arbitrary prime. The answer is analogous, but less simple :

**Theorem 5.8.** *Let $p$ be a prime number, and let $S$ be a $p$-Sylow subgroup of the finite group $G$.*

a) *Let $H$ be a subgroup of $S$ of index $p$ that satisfies the following stability property :*

(ST) *If $x \in S$ is $G$-conjugate to an element of $H$, then $x$ belongs to $H$. Define a map $f_H : G \to \{0, p\}$ by putting $f_H(x) = 0$ if the $p$-component*[5] *$x_p$ of $x$ is $G$-conjugate to an element of $H$, and $f_H(x) = p$ if not. Then $f_H$ is a virtual character of $G$ over $\mathbf{Q}$, with set of values $\{0, p\}$.*

b) *Conversely, every virtual character $f$ of $G$ over a field $K$ of characteristic 0, with set of values $\{0, p\}$, and with $f(1) = 0$, is equal to an $f_H$ as above, for a unique $H$.*

*Example.* Suppose that the $p$-Sylow subgroup $S$ is cyclic and non-trivial. In that case, $S$ has only one subgroup $H$ of index $p$, and hence there exists only one virtual character $f : G \to \{0, p\}$ with the required properties. If $m = |G|/p$, one may compute $f$ by the following rule : $f(x) = 0$ if $x^m = 1$, and $f(x) = p$ otherwise.

[5] Recall that every $x \in G$ can be written uniquely as $x = x'x_p = x_px'$, where $x_p$ has order a power of $p$ and $x'$ has order prime to $p$ ; the element $x_p$ is called the *$p$-component* of $x$.

*Remark.* When $p = 2$, every $H \subset S$ of index 2 with property (ST) is of the form $S \cap N$ where $N$ is a subgroup of index 2 of $G$; indeed the homomorphism $S \to \mathbf{Z}/2\mathbf{Z}$ with kernel $H$ is a "stable" element of the cohomology group $H^1(H, \mathbf{Z}/2\mathbf{Z})$, hence is the restriction of an element of $H^1(G, \mathbf{Z}/2\mathbf{Z})$, cf. [CE 56, Chap.XII, Theorem 10.1]. Thus, for $p = 2$, Theorem 5.8 is merely a reformulation of Theorem 5.2 and, in the proof below, we may assume $p > 2$.

*Proof of part* a) *of Theorem 5.8.*

We use induction on $|G|$. There are four steps :

i) *Proof when $S$ is cyclic of order $p$, is normal in $G$, and is its own centralizer in $G$.*

In that case $G$ is a semi-direct product $S.E$, where $e = |E|$ divides $p-1$, and the action of $E$ on $S$ is faithful (so that $G$ is a Frobenius group of order $ep$). The only possible $H$ is $H = 1$, and the corresponding function $f_H$ is equal to $p$ on $S - \{1\}$ and to 0 elsewhere. We have $f_H = p(1 - \mathrm{Ind}_E^G 1) + \frac{p-1}{e} r$, where $r$ is the character of the regular representation of $G$, and $\mathrm{Ind}_E^G 1$ is the character of the permutation representation of $G$ on $G/E$. This shows that $f_H$ is a virtual character of $G$ over $\mathbf{Q}$.

ii) *Proof when $S$ is normal in $G$.*

In that case, condition (ST) shows that $H$ is normal in $G$. The group $C = S/H$ is a $p$-Sylow subgroup of $G/H$, and is cyclic of order $p$. We may write $G/H$ as a semi-direct product $G/H = C.T$ and $T$ acts on $C$ by a homomorphism $\varepsilon : T \to \mathrm{Aut}(C) = \mathbf{F}_p^\times$. Let $E \subset \mathbf{F}_p^\times$ be the image of $\varepsilon$. We have a natural homomorphism $\pi : G \to G/H = C.T \to C.E$, and the function $f_H$ on $G$ is the inverse image by $\pi$ of the corresponding function on $C.E$; we then apply part i) above to $C.E$.

iii) *Proof when $G$ has a non-trivial normal subgroup $N$ of order prime to $p$.*

In that case, $f_H$ is the inverse image by $G \to G/N$ of an analogous function on $G/N$ and we apply the induction hypothesis to $G/N$.

iv) *Proof in the general case.*

By construction, $f_H$ is a class function on $G$ with values in $\mathbf{Q}$. To prove that it is a virtual character over $\mathbf{Q}$, it is enough to show that its restriction to every "$\Gamma_{\mathbf{Q}}$-elementary" subgroup has that property (Brauer-Witt's theorem, see e.g. [Se 78, §12.6, prop.36]). Let $G'$ be such a subgroup. We may conjugate $G'$ in such a way that $S' = S \cap G'$ is a $p$-Sylow subgroup of $G'$. Let $H'$ be $H \cap S' = H \cap G'$. The index of $H'$ in $S'$ is either 1 or $p$. In the first case, $f_H$ is 0 on $S'$, hence on $G'$, and there is nothing to prove. If $(S' : H') = p$, then the triple $(G', S', H')$ has property $(ST)$ and the corresponding map $f'_H : G' \to \{0,p\}$ is the restriction to $G'$ of $f_H$. We are thus reduced to the case where $G$ itself is $\Gamma_{\mathbf{Q}}$-elementary. This means that there exists a prime number $\ell$ and a normal cyclic subgroup $C$ of $G$, of order prime to $\ell$, such that $G/C$ is an $\ell$-group (such a group is often called "$\ell$-hyperelementary"). If $\ell = p$, this case follows from iii), applied with $N = C$ if $C \neq 1$ (if $C = 1$, then $G$ is a $p$-group and we apply ii)). If $\ell \neq p$, then $S$ is the unique $p$-Sylow subgroup of the cyclic group $C$, and we apply ii).

*Proof of part* b) *of Theorem 5.8 when $G$ is a $p$-group.*

We then have $G = S$. Let $f : G \to \{0,p\}$ be a virtual character of $G$ over a field $K$ of characteristic 0, with $f(1) = 0$. We want to prove that either $f = 0$ or $f$ is of the form $f_H$ for some $H \subset G$ of index $p$ (note that property (ST) is automatically satisfied here : every subgroup of index $p$ of a $p$-group is normal). We use induction on $|G|$. There are eight steps ; the first one does not require that $G$ is a $p$-group :

1) *If $x, y \in G$ generate the same cyclic subgroup, then $f(x) = f(y)$.*

This follows from the fact that the values of the virtual character $f$ belong to $\mathbf{Q}$.

[Recall the proof : we may write $y$ as $x^m$, where $m$ is prime to the order $N$ of $G$. Let $K_N$ be the field of the $N$-th roots of unity, and let $\sigma_m$ be the automorphism of $K_N$ that maps every $N$-th root of unity $z$ to its $m$-th power $z^m$. One has $f(x^m) = \sigma_m(f(x))$, cf. e.g. [Se 78, §13.1, Th.29] ; since $f(x)$ belongs to $\mathbf{Q}$, one has $\sigma_m(f(x)) = f(x)$, hence $f(x^m) = f(x)$.]

2) *If $f \neq 0$, the number of $x \in G$ with $f(x) = p$ is equal to $(1 - \frac{1}{p})|G|$.*

Write $|G|$ as $p^n$ and let $A$ be the set of $x \in G$ with $f(x) = p$. We have $\sum_{x \in G} f(x) = p|A|$. Since $f$ is a character, this sum is divisible by $p^n$, hence $|A|$ is divisible by $p^{n-1}$.

Let $\Delta$ be the unique subgroup of $(\mathbf{Z}/p^n\mathbf{Z})^\times$ of order $p - 1$. This group acts on the set $G$ by exponentiation : $x \mapsto x^\delta$. This action is free on $G - \{1\}$. By 1), $A$ is stable. Hence $|A|$ is divisible by $p - 1$.

We thus see that $|A|$ is divisible by $p^{n-1}(p-1)$, and, since $0 < |A| \leqslant p^n$, this implies $|A| = p^{n-1}(p-1) = (1 - \frac{1}{p})|G|$.

3) *If $f \neq 0$, and $G$ is cyclic, then $f(x) = 0 \iff x$ is a $p$-th power.*

If $G$ has order $p^n$, the set $C$ of generators of $G$ has order $p^{n-1}(p-1)$ and by 1), $f$ is constant on $C$. If $f$ were 0 on $C$, there would be at most $p^{n-1}$ elements $x$ with $f(x) = p$, which would contradict 2). Hence $f$ takes the value $p$ on $C$, and on no other element, because of 2).

4) *If $x \in G$ is a $p$-th power, then $f(x) = 0$.*

This follows from 3), applied to the subgroup of $G$ generated by an element $y$ such that $x = y^p$.

5) *Suppose $G$ is the direct product of two cyclic groups of the same order $p^n$, and $f \neq 0$. Then there exists a subgroup $H$ of $G$ of index $p$ such that $f = f_H$.*

Let $A$ be the set of $x \in G$ with $f(x) = p$, and let $B = G - G^p$ be the set of elements of $G$ that are not $p$-th powers. By 4), we have $A \subset B$, and the inclusion is strict because of 2). Choose $x \in B - A$, and let $H$ be a subgroup of $G$ of index $p$ containing $x$. Suppose that the restriction $f|H$ of $f$ to $H$ is not 0. We then have $|A \cap H| = p^{2n-2}(p-1)$ by vi). But $H$ contains $G^p$ that has order $p^{2n-2}$ and on which $f$ vanishes by 4). This shows that $A \cap H = H - G^p$, contrary to the fact that $x \notin A$. Hence $f$ is 0 on $H$, and, by 2), it is equal to $p$ on $G - H$.

6) *Let $x, y \in G$ with $xy = yx$ and $f(x) = f(y) = 0$. Then $f(xy) = 0$.*

We may assume that $G$ is generated by $x$ and $y$, hence is a quotient of a group of type 5). The conclusion follows from 5).

7) *Let $C$ be a central subgroup of $G$ of order $p$ such that $f|C = 0$. Then $f$ is of the required form $f_H$.*

Indeed, by 6), $f$ is constant mod $C$, and one applies the induction assumption to $G/C$.

8) *End of the proof of* b) *when $G$ is a $p$-group.*

The cases where $|G| = 1$ or $p$ are trivial. The case $|G| = p^2$ is a consequence of 5), since $G$ is commutative and can be generated by two elements. We may thus assume that $|G| > p^2$. By 7), it is enough to prove that there exists a central subgroup $C$ of $G$, of order $p$, such that $f|C = 0$. Choose first any central subgroup $C_0$ of $G$ of order $p$. If $f|C_0 = 0$, we take $C = C_0$. If not, choose $H \subset G$ of index $p$ with $C_0 \subset H$ : this is possible because $|G| > p$. We have $f|H \neq 0$, and the induction assumption, applied to $H$, shows that the elements $x \in H$ with $f(x) = 0$ make up a subgroup $H'$ of $H$ of index $p$. Since $C_0$ is not contained in $H'$, we have $H = C_0 \times H'$, and $H'$ is non-trivial because $|G| > p^2$. Moreover, $H'$ is normal in $G$ because $f$ is a class function. Hence the intersection of $H'$ with the center $Z(G)$ of $G$ is non-trivial. If $C$ is a subgroup of order $p$ of $H' \cap Z(G)$, $C$ is central in $G$ and $f|C = 0$.

*Proof of part* b) *of Theorem 5.8 in the general case.*

Let $f : G \to \{0, p\}$ be a virtual character, with $f(1) = 0$. We show first :

9) *Let $x \in G$ and let $x_p$ be its $p$-component. We have $f(x) = f(x_p)$.*

One may write $x_p$ as $x^m$, for some integer $m$ prime to $p$. Hence it is enough to prove that $f(x^\ell) = f(x)$ for every prime $\ell \neq p$. This follows from the congruence $f(x^\ell) \equiv f(x) \pmod{\ell}$, that is valid for every virtual character with values in $\mathbf{Z}$. [More generally, if $f$ is any virtual character with values in the ring of integers $A$ of a number field, and if $\mathfrak{p}$ is any prime ideal of $A$ dividing $\ell$, one has $f(x^\ell) \equiv f(x)^\ell \bmod \mathfrak{p}$.]

With the notation of 9), we have $f(x) = f(y)$ for every $y \in S$ that is $G$-conjugate to $x_p$. This shows that $f$ is determined by its restriction $f|S$ to $S$; since we already know (see above) the structure of $f|S$, this concludes the proof.

*Exercises.*

1) Let $G = A_5$ be the alternating group on 5 elements. The group $G$ can act transitively on 5 or 6 elements; let $1 + \chi_4$ and $1 + \chi_5$ be the corresponding permutation characters, where $\chi_4$ and $\chi_5$ are irreducible. Check that the set of values of $1 + \chi_4 - \chi_5$ is $\{0, 3\}$.

2) Let $G = \mathbf{PSL}_2(\mathbf{F}_q)$, and let $\varphi$ be the Steinberg character of $G$; its degree is $q$.

Suppose that $q \equiv 1 \pmod 3$, and let $\varepsilon$ be a character of order 3 of $\mathbf{F}_q^\times$. Let $\psi$ be the irreducible character of the principal series of $G$ associated with $\varepsilon$; its degree is $q+1$. Show that the set of values of $1 + \varphi - \psi$ is $\{0, 3\}$.

Suppose that $q \equiv -1 \pmod 3$. Let $\varepsilon'$ be a character of order 3 of $\mathbf{F}_{q^2}^\times$ and let $\psi'$ be the irreducible character of the complementary series of $G$ associated with $\varepsilon'$; the degree of $\psi'$ is $q-1$. Show that the set of values of $1 + \psi' - \varphi$ is $\{0, 3\}$.

When $q = 4$ or 5, one has $G \simeq A_5$ and one recovers (in two different ways) the example of Exerc.1.

3) Show that the symmetric group $S_n$ has a virtual character with set of values $\{0, p\}$ if and only if its $p$-Sylow subgroups have order $p$, i.e., if $p \leqslant n < 2p$.

## 5.2. Density estimates

### 5.2.1. Definitions

Let $G$ be a compact topological group (resp. an algebraic group) and let $Z$ be a closed subset (resp. a closed subscheme) of $G$. We shall define the *Haar density* (resp. the *Zariski density*) of $Z$ in $G$ as follows :

5.2.1.1. Suppose first that $G$ is a compact group. We put on $G$ its Haar measure $\mu$ of total mass 1 ("normalized Haar measure"), and we define the Haar density of $Z$, denoted by $\mathrm{dens}_G^{\mathrm{haar}}(Z)$, as the measure $\mu(Z)$ of $Z$; this makes sense since every closed subset is measurable.

When $G$ is profinite, there is a simple formula for $\mathrm{dens}_G^{\mathrm{haar}}(Z)$, namely :

$$\mathrm{dens}_G^{\mathrm{haar}}(Z) = \mathrm{inf}_U |Z_U|/(G:U),$$

where $U$ runs through the normal open subgroups of $G$ and $Z_U$ denotes the image of $Z$ in the finite group $G/U$. This follows for instance from [INT, Chap.IV, §1, Th.1], applied to the characteristic function of the open set $G-Z$; one may also take this formula as the *definition* of $\mu(Z)$; this is essentially what is done in [FJ 08, Chap.18].

5.2.1.2. Suppose that $G$ is an algebraic group over a field $F$, and that $Z$ is a closed subscheme of $G$. Let $n_G$ be the number of geometric connected components[6] of $G$ and let $n_G(Z)$ be the number of these components that are contained in $Z$. We define the Zariski density of $Z$, denoted by $\mathrm{dens}_G^{\mathrm{zar}}(Z)$, as the quotient $n_G(Z)/n_G$. For instance, every $Z$ of dimension $< \dim G$ has density 0; if $G$ is connected, $\mathrm{dens}_G^{\mathrm{zar}}(Z) = 0$ for every $Z \neq G$, and $\mathrm{dens}_G^{\mathrm{zar}}(Z) = 1$ for $Z = G$.

*Remark.* These density notions are *compatible with passage to quotient* in the following sense : suppose first that we are in the topological setting of §5.2.1.1, and that $N$ is a normal closed subgroup of $G$ such that $N.Z = Z$, so that $Z$ is the inverse image by $G \to G/N$ of a closed subset $Z_{G/N}$ of $G/N$. Then $\mathrm{dens}_G^{\mathrm{haar}}(Z) = \mathrm{dens}_{G/N}^{\mathrm{haar}}(Z_{G/N})$; this follows from the fact that the Haar measure of $G/N$ is the image [7] of the Haar measure of $G$ by the map $G \to G/N$. There is a similar result for the Zariski density.

## 5.2.1. Densities in compact Lie groups

Let $K$ be a locally compact non-discrete field of characteristic 0, i.e. either $\mathbf{R}$, $\mathbf{C}$ or a finite extension of an $\ell$-adic field $\mathbf{Q}_\ell$, cf. [AC V-VI, Chap.VI, §9, Th.1].

Let $G$ be a compact analytic Lie group over $K$ of finite dimension; recall, cf. [LIE III, §1], that this means that the group $G$ is endowed with a structure of analytic $K$-manifold that is compatible with its group structure. As in §5.2.1.1, we denote by $\mu$ the normalized Haar measure of $G$; it is well known ([LIE III, §9.16]) that $\mu$ is the measure associated with a left and right non-zero invariant differential form of degree equal to $\dim_K G$.

**Proposition 5.9.** *Let $Z$ be a closed $K$-analytic subset[8] of $G$.*

---

[6] We mean by this the connected components of $G_{\overline{F}}$, where $\overline{F}$ is an algebraic closure of $F$.

[7] Recall that, if $f : X \to Y$ is a continuous map of compact spaces, and if $\mu$ is a measure on $X$, the image $f_*(\mu)$ of $\mu$ by $f$ is the unique measure $\nu$ on $Y$ such that $\nu(\varphi) = \mu(\varphi \circ f)$ for every continuous function $\varphi$ on $Y$.

[8] A closed subset of an analytic manifold is called *analytic* if it is defined locally by the vanishing of a finite number of analytic functions.

1) *The set $Z$ is a disjoint union $Z = Z_0 \cup Z_1$, where $Z_1$ is open and closed, and $Z_0$ is analytic, closed and of empty interior. Such a decomposition is unique.*

2) *One has $\mu(Z_0) = 0$.*

*Proof.* Let $Z_1$ be the interior of $Z$; since $Z$ is analytic, $Z_1$ is closed (this is the precise form of the classical "principle of analytic continuation"); hence $Z_1$ is both open and closed. One then defines $Z_0$ as the complement of $Z_1$ in $Z$, i.e. as the boundary of $Z$. The decomposition of $Z$ so obtained is obviously unique. This proves 1).

Assertion 2) is well-known. It amounts to saying that, if $f$ is a non-zero power series converging in an open polydisk $U$ of $K^n$, then the set $Z_f$ of zeros of $f$ in $U$ has measure 0 in a neighbourhood of 0; this is proved by choosing the local coordinates in such a way that each fiber of the projection $Z_f \to K^{n-1}$ is finite, hence has measure 0 in $K$; by Fubini's theorem, this implies that $Z_f$ has measure 0.

**Corollary 5.10.** *One has $\mu(Z) = 0$ if and only if the interior $Z_1$ of $Z$ is empty.*

This follows from the fact that the Haar measure of a non-empty open set is $> 0$.

**Corollary 5.11.** *There exists an open normal subgroup $N$ of $G$ such that $Z_1$ is stable under left and right multiplication by $N$; for such $N$, the density of $Z$ is an integral multiple of $1/(G:N)$.*

*Proof.* Let $H$ be the set of all $g \in G$ such that $gZ_1 = Z_1$. Since $Z_1$ is open and closed, the same is true for $H$, which is thus an open subgroup of finite index of $G$. By taking the intersection of its conjugates, we get an open normal subgroup $N$ of $G$ such that $NZ_1 = Z_1 = Z_1N$. In other words, $Z_1$ is a union of $N$-cosets of $G$; if $h$ is the number of these cosets, we have $\mathrm{dens}^{\mathrm{haar}}(Z) = \mu(Z_1) = h.\mu(N) = h/(G:N)$.

### 5.2.2. Comparing Haar density with Zariski density

We are now going to relate the two kinds of "density" defined in §5.2.1. Let $K$ be as in §5.2.2 above.

**Proposition 5.12.** *Let $H$ be an algebraic group over $K$ and let $G$ be a compact subgroup of the $K$-Lie group $H(K)$. Assume that $G$ is Zariski-dense in $H$. Let $R$ be a closed subscheme of $H$. Put $R_G = G \cap R(K)$. We have*

$$\mathrm{dens}_G^{\mathrm{haar}}(R_G) = \mathrm{dens}_H^{\mathrm{zar}}(R).$$

[Hence the Haar density of a subset of $G$ defined by algebraic equations can be computed algebraically.]

*Proof.* Let $H^0$ be the identity component of $H$. Since $G$ is Zariski-dense in $H$, every connected component of $H$ (over an algebraic closure $\overline{K}$ of $K$) meets $G$. If we put $G^0 = G \cap H^0(K)$, the map $G/G^0 \to H/H^0$ is a bijection; put $m = |G/G^0|$.

Let $V$ be a $G^0$-coset of $G$ and let $\overline{V}$ be its Zariski closure, which is a connected component of $H$. We may assume that $R$ is contained in $\overline{V}$ : the general case will follow by additivity. There are two cases :

a) $R = \overline{V}$. In that case $R_G$ is equal to $V$ and we have

$$\mathrm{dens}^{\mathrm{haar}}(R_G) = \frac{1}{m} = \mathrm{dens}^{\mathrm{zar}}(R).$$

b) $R \neq \overline{V}$. In that case we have $\mathrm{dens}^{\mathrm{zar}}(R) = 0$ and we need to show that the same formula holds for the Haar density of $R_G$ in $G$. Let $K_0$ be the closure of $\mathbf{Q}$ in $K$, namely :

$K_0 = \mathbf{R}$ if $K = \mathbf{R}$ or $\mathbf{C}$;

$K_0 = \mathbf{Q}_\ell$ if $K$ is a finite extension of $\mathbf{Q}_\ell$.

Note that $H(K)$ has a natural structure of $K_0$-Lie group (since $K$ is a finite extension of $K_0$) ; since $G$ is a closed subgroup of that group, it is a $K_0$-Lie subgroup ([LIE III, §8, Th.2] - this is where the fact that $K_0 = \mathbf{R}$ or $\mathbf{Q}_\ell$ is used). The set $R_G$ is $K$-analytic, hence also $K_0$-analytic and we may apply cor.5.10 to it (over the ground field $K_0$) provided we prove that the interior $U$ of $R_G$ is empty. Suppose it is not. Since $U$ is open and closed there exists an open normal subgroup $N$ of $G^0$ such that $NU = U$. Let $\overline{N}$ be the Zariski closure of $N$ in $H$. It is a subgroup of finite index of $H^0$ ; since $H^0$ is connected, we have $\overline{N} = H^0$. Let $u$ be a point of $U$. We have $Nu \subset U \subset R(K)$, hence $\overline{Nu} \subset R$, which shows that $R$ contains $\overline{Nu} = H^0 u = \overline{V}$, contrary to what we had assumed. This concludes the proof.

## 5.3. The unitary trick

### 5.3.1. The discrete case

Let $G$ be a group, let $K$ be a field of characteristic 0, and let

$$\rho : G \to \mathbf{GL}_n(K)$$

be a linear representation, that we assume to be semisimple.

To these data, we are going to associate a compact subgroup $G_c$ of $\mathbf{GL}_n(\mathbf{C})$. Such a group may be viewed as a *unitary analog* of $(G, \rho)$. Its construction is made in four steps :

1) Let $H = \overline{G}$ be the Zariski closure of $\rho(G)$ in the $K$-variety $\mathbf{GL}_{n/K}$. Since $\rho$ is semisimple, the same is true for the natural $n$-dimensional representation of $H$; this implies that $H$ is a reductive (not necessarily connected) algebraic subgroup of $\mathbf{GL}_{n/K}$.

2) Choose a subfield $K_1$ of $K$ that is embeddable in $\mathbf{C}$ (e.g. finitely generated over $\mathbf{Q}$), and over which $H$ can be defined, so that we may view $H$ as a $K_1$-subgroup scheme $H_1$ of $\mathbf{GL}_{n/K_1}$ and we have $\rho(G) \subset H_1(K_1)$.

3) Choose an embedding $\iota : K_1 \to \mathbf{C}$. Let $H_\iota$ be the $\mathbf{C}$-algebraic subgroup of $\mathbf{GL}_{n/\mathbf{C}}$ deduced from $H_1$ by the base change $\iota$. We have a natural embedding

$$\rho(G) \to H_1(K_1) \overset{\iota}{\to} H_\iota(\mathbf{C}),$$

and $\iota(\rho(G))$ is Zariski-dense in $H_\iota(\mathbf{C})$.

4) We now choose for $G_c$ any maximal compact subgroup of the complex Lie group $H_\iota(\mathbf{C})$. Note that $G_c$ is Zariski-dense in $H_\iota$, see e.g. [Se 93, §5.3, Theorem 4].

*Remark.* The construction of $G_c$ depends on several choices, the most important one being that of $\iota$ in step 3); hence there is no uniqueness (for an explicit example, see the Exercise below).

**Proposition 5.13.** *Let $R$ be a closed subscheme of* $\mathbf{GL}_n$ *that is defined over* $\mathbf{Q}$. *We have* $\rho(G) \subset R(K) \iff G_c \subset R(\mathbf{C})$.
[In other words : the $\mathbf{Q}$-equations satisfied by $\rho(G)$ and by $G_c$ are the same.]

*Proof.*

We have $\rho(G) \subset R(K) \iff H \subset R$ since $\rho(G)$ is Zariski-dense in $H$.

We have $H \subset R \iff H_\iota \subset R$ since $R$ is defined over $\mathbf{Q}$ and $\iota$ is the identity on $\mathbf{Q}$.

We have $H_\iota \subset R \iff G_c \subset R(\mathbf{C})$ since $G_c$ is Zariski-dense in $H_\iota(\mathbf{C})$.

*Example.* If $\rho(G)$ is contained in $\mathbf{GL}_a(K) \times \mathbf{GL}_b(K)$, with $n = a+b$, then $G_c$ is contained in $\mathbf{GL}_a(\mathbf{C}) \times \mathbf{GL}_b(\mathbf{C})$, and conversely.]

*Exercise.* (Construction of a reductive group $H$ over $\mathbf{C}$ and of an automorphism $\iota$ of $\mathbf{C}$ such that the group $H_\iota$ is not isomorphic to $H$.)

a) Let $A = \mathbf{PSL}_5(\mathbf{F}_{11})$ and let $\tilde{A} = \mathbf{SL}_5(\mathbf{F}_{11})$. The kernel $C$ of the natural projection $\tilde{A} \to A$ is cyclic of order 5; it is the unique cyclic subgroup of $\mathbf{F}_{11}^\times$ of order 5, i.e. the group generated by the homothety $c$ of ratio 3. The group $C$ is the Schur multiplier of $A$. The group $\mathrm{Aut}(A) = \mathrm{Aut}(\tilde{A})$ is a dihedral group of order 10, which acts on $C$ by $c \mapsto c^{\pm 1}$.

b) Let $\mu_5$ be the group of 5-th roots of unity, viewed as a subgroup of the multiplicative group $\mathbf{G}_m = \mathbf{GL}_1$ . If $z \neq 1$ belongs to $\mu_5$, define an algebraic group $H_z$ as the quotient of $\mathbf{G}_m \times \tilde{A}$ by the group of order 5 generated by $(z, c)$. We have an exact sequence $1 \to \mathbf{G}_m \to H_z \to A \to 1$.

Let $z'$ be another choice of $z$ in $\mu_5$. Show that the algebraic groups $H_z$ and $H'_z$ are isomorphic if and only if $z' = z^{\pm 1}$.

[Hint. Use the fact that both $\mathrm{Aut}(\mathbf{G}_m)$ and $\mathrm{Aut}(A)$ act on $C = \mathbf{G}_m \cap \tilde{A}$ by $c \mapsto c^{\pm 1}$.]

c) Let $\iota$ be an automorphism of $\mathbf{C}$ which transforms $\sqrt{5}$ into $-\sqrt{5}$. Show that $\iota(z) = z^{\pm 2}$; hence the group $H = H_z$ is not isomorphic to its $\iota$-transform $H_\iota$. (Note that $H$ is not connected. Indeed, if $G$ is any connected reductive group over an algebraically closed field $k$, and if $\iota$ is any endomorphism of $k$, the group $G_\iota$ is isomorphic to $G$; this follows from the classification theorem of such groups via their *données radicielles*, cf. [SGA 3, vol.3, exposé XXIII, §5].)

### 5.3.2. The compact case

We keep the notation and the hypotheses of the previous section, to which we add the following assumptions :

(i) $G$ is a compact topological group.

(ii) $K$ is a locally compact non-discrete topological field.

(iii) $\rho : G \to \mathbf{GL}_n(K)$ is continuous.

[Note that the cardinality of $K$ is $2^{\aleph_0}$, hence we may[9] take $K_1 = K$ in the second step of the construction of $G_c$.]

We have the following improvement of Proposition 5.13 :

**Theorem 5.14.** *Let $R$ be a closed subscheme of $\mathbf{GL}_n$. Assume that $R$ is defined over $\mathbf{Q}$. Let $R_G$ be the set of $g \in G$ such that $\rho(g) \in R(K)$, and let $R_{G_c} = G_c \cap R(\mathbf{C})$. We have* $\mathrm{dens}_G(R_G) = \mathrm{dens}_{G_c}(R_{G_c})$.

[Here both densities are Haar densities, cf. §5.2.1.1. What the theorem says is that, in order to compute such a density in $G$, it is enough to compute it in its unitary analog $G_c$, provided that the defining equations of $R$ have coefficients in $\mathbf{Q}$.]

*Proof.* Let us write $H = \overline{G}$, as above, and let $R_{\rho(G)} = \rho(G) \cap R(K)$. We follow the same pattern as in the proof of Proposition 5.13, namely we prove the following equalities :

$$\begin{aligned}\mathrm{dens}_G(R_G) = \mathrm{dens}_{\rho(G)}(R_{\rho(G)}) &= \mathrm{dens}^{\mathrm{zar}}_H(R \cap H) = \mathrm{dens}^{\mathrm{zar}}_{H_\iota}(R \cap H_\iota) \\ &= \mathrm{dens}_{G_c}(R_{G_c}).\end{aligned}$$

The first equality (on the left) is true because $R_G$ is the inverse image of $R_{\rho(G)}$ by $G \to \rho(G)$, cf. §5.2.1.

[9] As Deligne says in [DE 80, 1.2.11], this is just a *commodité d'exposition.* A down-to-earth reader will prefer to choose $K_1$ finitely generated over $\mathbf{Q}$.

The second one follows from Proposition 5.12 applied to $\rho(G)$, $H$ and $R \cap H$.

The third one is true because $R$ is defined over $\mathbf{Q}$.

The fourth one follows from Proposition 5.12 applied (over the field $\mathbf{C}$) to $G_c$, $H_\iota$ and $R \cap H_\iota$.

*Remark.* Note the auxiliary, but essential, role of the Zariski density in the above proof; its purely algebraic definition makes it compatible with such strange-looking field embeddings as $\iota : K \to \mathbf{C}$, where $K$ is an $\ell$-adic field.

### 5.3.3. An example

Let $G$ and $K$ be as above, namely :

- $G$ is a compact topological group,
- $K$ is a locally compact non-discrete topological field of characteristic 0.

[In the next chapter, $G$ will be a profinite group and $K$ an $\ell$-adic field.]

Consider two continuous linear representations

$$\rho_1 : G \to \mathbf{GL}_a(K) \ \text{ and } \ \rho_2 : G \to \mathbf{GL}_b(K),$$

where $a$ and $b$ are $\geqslant 0$ and not both 0. Let $\chi_i$ be the character of $\rho_i$, and define $f \in \mathrm{Cl}(G, K)$ as $f = \chi_1 - \chi_2$. We are going to show that, if $f$ is not everywhere 0, the subset of $G$ where it is $\neq 0$ has a density that is larger than some constant depending only on $a$ and $b$. More precisely :

**Theorem 5.15.** *If $f = \chi_1 - \chi_2$ is not identically 0, the set $I_f$ of the points $g \in G$ with $f(g) \neq 0$ has density $\geqslant \frac{1}{N}$ , where $N = (a+b).\sup(a, b)$.*

When $b = 0$, this gives :

**Corollary 5.16.** *The set of points $g \in G$ with $\chi_1(g) \neq 0$ has density $\geqslant 1/a^2$.*

This bound is optimal : for every $a \geqslant 1$ there exists a finite group $G$ and an irreducible representation $\rho$ of $G$ of degree $a$, such that the set of $g \in G$ with $\mathrm{Tr}(\rho(g)) \neq 0$ has density $1/a^2$.

[Hint. Take for $G$ a Heisenberg-like group of order $a^3$, cf. [Se 81, p.172].]

*Proof of Theorem 5.15 when* $K = \mathbf{C}$

We may assume $a \geqslant b$ and also that the two representations $\rho_1$ and $\rho_2$ do not contain a common irreducible subrepresentation. We need to give a lower bound for $\mu(I_f)$, under the assumption that $\mu(I_f) > 0$.

Since $\chi_1$ is a non-zero character, the orthogonality relations of characters imply that $\int_G |\chi_1|^2 \geqslant 1$, the integral being taken with respect to the normalized Haar measure $\mu$ of $G$; they also imply that the characters $\chi_1$ and $\chi_2$ are orthogonal. We then have :

$$1 \leqslant \int_G |\chi_1|^2 = \int_G \overline{\chi}_1.(\chi_1 - \chi_2) = \int_{I_f} \overline{\chi}_1.(\chi_1 - \chi_2) \leqslant \mu(I_f).a(a+b),$$

since $|\chi_1| \leqslant a$ and $|\chi_1 - \chi_2| \leqslant a+b$. This shows that $\mu(I_f) \geqslant 1/a(a+b)$, as claimed.

*Proof of Theorem 5.15 in the general case*

We may assume that the $\rho_i$ are semisimple, since the traces do not change under semisimplification. We also assume that $f$ is not 0, i.e. that $\rho_1$ and $\rho_2$ are not isomorphic.

Put $n = a+b$ and embed $\mathbf{GL}_a \times \mathbf{GL}_b$ in $\mathbf{GL}_n$ in a standard way (by splitting the interval $[1, n]$ as $[1, a] \sqcup [a+1, a+b]$, say). The pair $(\rho_1, \rho_2)$ thus defines a linear representation

$$\rho : G \rightarrow \mathbf{GL}_a(K) \times \mathbf{GL}_b(K) \rightarrow \mathbf{GL}_n(K).$$

By applying to $\rho$ the construction of §5.3.1 and §5.3.2 we get a compact subgroup $G_c$ of $\mathbf{GL}_n(\mathbf{C})$. Proposition 5.13 shows that $G_c$ is contained in $\mathbf{GL}_a(\mathbf{C}) \times \mathbf{GL}_b(\mathbf{C})$. Let $R$ be the closed subscheme of $\mathbf{GL}_n$ made up of the points $(g_1, g_2) \in \mathbf{GL}_a \times \mathbf{GL}_b$ such that $\mathrm{Tr}(g_1) = \mathrm{Tr}(g_2)$. By Theorem 5.14, the sets $R_G$ and $R_{G_c}$ have the same density. But we have just shown that the complement of $R_{G_c}$ in $G_c$ (which we denoted by $I_f$) has density $\geqslant 1/a(a+b)$. Hence we get the same bound for $R$ : its complement in $G$ has density $\geqslant 1/a(a+b)$.

*Remark.* As mentioned above, the bound $1/a(a+b)$ is sharp when $b = 0$. It is also sharp when $b = a$, cf. exerc. 1. In most other cases, it is not sharp, cf. exerc. 2.

*Problem.* One may wonder whether Corollary 5.16 extends to characteristic $p > 0$. More precisely, let $G$ be a finite subgroup of $\mathbf{GL}_a(k)$, with char $k > 0$, and let $\varepsilon$ be the density of the set of $g \in G$ with $\mathrm{Tr}(g) \neq 0$. Is it true that $\varepsilon$ is either 0 or $\geqslant 1/a^2$ ? This does not look likely, but I do not know any counterexample.

*Exercises.*

1) Let $G$ be a finite group and let $\rho : G \rightarrow \mathbf{GL}_a(\mathbf{C})$ be a faithful irreducible representation with the property that its character $\chi$ vanishes outside the center $C$ of $G$. Such a pair $(G, \rho)$ exists for every $a \geqslant 1$, and one has $(G : C) = a^2$, cf. [Se 81, p.172].

Let $G_1 = G \times \{\pm 1\}$; let $\psi : G_1 \rightarrow \{\pm 1\}$ be the second projection, and let us view $\chi$ as a character of $G_1$ via the first projection $G_1 \rightarrow G$. Let $f = \chi - \psi\chi$.

Show that the density of the set of elements $z$ of $G_1$ with $f(z) \neq 0$ is $1/2a^2$ (hence the bound of Theorem 5.15 is sharp when $a = b$).

2) With the notation of Theorem 5.15, suppose that $a > b$. Show that the bound $1/a(a+b)$ can be improved to $1/(a-b)^2$ if $b \leqslant a/3$, and to $1/b(a+b)$ if $b \geqslant a/3$. When $(a, b) = (2, 1)$, the bound so obtained is $1/3$; show that the optimal bound is $2/3$, and that it is attained by taking for $G$ the binary tetrahedral group $\tilde{A}_4 \simeq \mathbf{SL}_2(\mathbf{F}_3)$.

3) Let $f = \chi_1 - \chi_2$ be as in Theorem 5.15. Let $F$ be the virtual character

$$F = \lambda^2 f = \lambda^2 \chi_1 + \lambda^2 \chi_2 - \chi_1 \chi_2.$$

Show that, either $F = 0$, or the set of points $g \in G$ with $F(g) \neq 0$ has density $\geqslant 1/C^2$ where $C = (a+b)(a+b-1)/2$.

[Hint. Same method as for Theorem 5.15 : reduction to the case where the ground field $K$ is $\mathbf{C}$, in which case the result follows from the bound $|F| \leqslant C$.]

CHAPTER 6

# THE $\ell$-ADIC PROPERTIES OF $N_X(p)$

In this chapter, the ground field is $\mathbf{Q}$ ; the set of prime numbers is denoted by $P$. We put, as usual, $\Gamma_{\mathbf{Q}} = \mathrm{Gal}(\overline{\mathbf{Q}}/\mathbf{Q})$ ; if $S$ is a finite subset of $P$, we denote by $\Gamma_S$ the largest quotient of $\Gamma_{\mathbf{Q}}$ that is unramified outside $S$, i.e. the fundamental group of $\mathrm{Spec}\,\mathbf{Z} - S$ relative to the geometric point $\mathbf{Z} \to \overline{\mathbf{Q}}$. If $p \notin S$, we denote by $\sigma_p$ the corresponding Frobenius element of $\Gamma_S$ ; it is well defined up to conjugation ; its inverse $\sigma_p^{-1}$ (the "geometric Frobenius") will be denoted by $g_p$.

## 6.1. $N_X(p)$ viewed as an $\ell$-adic character

### 6.1.1. The Galois character given by cohomology

Let $X$ be a scheme of finite type over $\mathbf{Z}$; to simplify the notation, we denote by $X_0$ the corresponding $\mathbf{Q}$-variety (i.e. the generic fiber $X_{/\mathbf{Q}}$ of $X \to \mathrm{Spec}\,\mathbf{Z}$), and let $\overline{X}$ be the $\overline{\mathbf{Q}}$-variety obtained from $X_0$ by the base change $\mathbf{Q} \to \overline{\mathbf{Q}}$.

*Let us first assume that the scheme $X$ is separated* (see below for the general case). If $\ell$ is a prime number and $i$ is an integer $\geqslant 0$, $H^i_c(\overline{X}, \mathbf{Q}_\ell)$ is the $i$-th $\ell$-adic cohomology group of $\overline{X}$ with proper support, cf. §4.1 ; recall that it is a finite-dimensional $\mathbf{Q}_\ell$-vector space, that vanishes for $i > 2\dim X_0$. In order to simplify the notation, we shall write it $H^i(X, \ell)$.

There is a natural action of $\Gamma_{\mathbf{Q}} = \mathrm{Gal}(\overline{\mathbf{Q}}/\mathbf{Q})$ on each $H^i(X, \ell)$. By Theorem 4.13, applied to $K = \mathbf{Q}$, there exists a finite subset $S$ of $P$, containing $\ell$, such that, if $p \notin S$, the action of $\Gamma_{\mathbf{Q}}$ on every $H^i(X, \ell)$ is unramified outside $S$; this gives an action of $\Gamma_S$ on $H^i(X, \ell)$. If $p \notin S$, let us denote by $\mathrm{Tr}(g_p | H^i(X, \ell))$ the trace of the geometric Frobenius $g_p$ acting on $H^i(X, \ell)$, and define similarly $\mathrm{Tr}(g_p^e | H^i(X, \ell))$ for every $e \geqslant 1$. By Theorem 4.13, we have :

**Theorem 6.1.** *There exists a choice of $S$ such that*

$$N_X(p^e) = \sum_i (-1)^i \mathrm{Tr}(g_p^e | H^i(X, \ell)) \quad \textit{for every } p \notin S \textit{ and every } e \geqslant 1.$$

[This formula extends to every $e \in \mathbf{Z}$, provided $N_X(p^e)$ is defined as explained in §1.5. The proof is the same.]

We may reformulate this in the style of Chap.V, by introducing the character $h_{i,X,\ell}$ of the action of $\Gamma_S$ on $H^i(X,\ell)$, together with the virtual character

$$h_{X,\ell} = \sum_i (-1)^i h_{i,X,\ell}.$$

Theorem 6.1 then means that

$$N_X(p) = h_{X,\ell}(g_p) \text{ for every } p \notin S,$$

and more generally

$$N_X(p^e) = h_{X,\ell}(g_p^e) \quad \text{for every } p \notin S \text{ and every } e \geqslant 1.$$

When there is no possible confusion about $X$ and $\ell$, we shall write $h$ instead of $h_{X,\ell}$ and $h_i$ instead of $h_{i,X,\ell}$.

*Remarks.*

1. The Galois representations $H^i(X,\ell)$ depend only on $X_0$ and $\ell$. The set $S$ depends on $\ell$ (we shall sometimes write it $S_\ell$) and on $X$ (and not merely on $X_0$); indeed, if one makes a **Q**-change of coordinates, this modifies $N_X(p^e)$ for a finite set of $p$ and one has to change $S$ accordingly.

2. The formula $N_X(p) = h(g_p)$ shows that the values of the virtual character $h$ on the geometric Frobenius elements belong to **Z** and are independent of $\ell$, provided $p$ is large enough and distinct from $\ell$. It is conjectured that the same is true for each character $h_i$.

3. We shall see in the next chapter that there is a different decomposition of $h_X$ as a sum of virtual characters, which is better suited for archimedean estimates; it is called the *weight* decomposition.

4. Some readers may prefer to express Theorem 6.1 with the arithmetic Frobenius $\sigma_p$ instead of the geometric one. This is easy; all one has to do is to define the *homology group* $H_i(X,\ell)$ as the $\mathbf{Q}_\ell$-dual of $H^i(X,\ell)$; the group $\Gamma_{\mathbf{Q}}$ acts on $H_i(X,\ell)$ and Theorem 6.1 may be reformulated as :

$$N_X(p^e) = \sum_i (-1)^i \operatorname{Tr}(\sigma_p^e | H_i(X,\ell)) \quad \textit{for every } p \notin S_\ell \textit{ and every } e \geqslant 1.$$

[Indeed, if $\sigma$ is an automorphism of a finite dimensional vector space $V$, and if $\sigma^*$ is the corresponding automorphism of the dual $V^*$ of $V$, one has $\operatorname{Tr}(\sigma^*|V^*) = \operatorname{Tr}(\sigma^{-1}|V)$, since $\sigma^*$ is the transpose inverse ${}^t\sigma^{-1}$ of $\sigma$.]

One of the most important properties of $h_{X,\ell}$ is that it depends additively on $X$ (and even on $X_0$) :

**Theorem 6.2.**

a) *Let* $F$ *be a closed subscheme of* $X$ *and let* $U = X - F$. *Then* $h_{X,\ell} = h_{F,\ell} + h_{U,\ell}$.

b) *Let* $(U_i)_{i\in I}$ *be a finite open covering of* $X$. *If* $J$ *is any subset of* $I$, *put* $U_J = \bigcap_{i\in J} U_i$. *Then* $\sum_{J\subset I}(-1)^{|J|}h_{U_J,\ell} = 0$.

[Another way to write b) is : $h_{X,\ell} = \sum_{J\subset I, J\neq\varnothing}(-1)^{|J|+1}h_{U_J,\ell}$.]

*Proof of* a). We have $N_X(p) = N_F(p) + N_U(p)$ for every $p$. By Theorem 6.1, this implies that there exists a finite set $S$ of primes such that the two virtual characters $h_{X,\ell}$ and $h_{F,\ell} + h_{U,\ell}$ take the same value on all the $g_p$, for $p \notin S$. Since the $g_p$ are dense in $\mathrm{Cl}\,\Gamma_S$, these characters are equal.

*Alternative method.* Use the exact sequence connecting the cohomology groups $H^i(X,\ell)$, $H^i(F,\ell)$ and $H^i(U,\ell)$, cf. §4.1.

*Proof of* b). Same as the proof of a), using the combinatorial identity

$$\sum_{J\subset I}(-1)^{|J|}N_{U_J}(p) = 0.$$

*How to remove the assumption that the scheme* $X$ *is separated over* Spec $\mathbf{Z}$.

The only reason we asked that $X$ is separated was to insure that its cohomology with proper support is well defined. This rather artificial condition can be dispensed with, at the cost of cutting up $X$ into smaller pieces. For instance, one may choose a finite open covering $(U_i)_{i\in I}$ of $X$ where all the $U_i$ are separated (e.g. affine), and define $h_{X,\ell}$ by the formula given above, namely

$$h_{X,\ell} = \sum_{J\subset I, J\neq\varnothing}(-1)^{|J|+1}h_{U_J,\ell}, \quad \text{where} \quad U_J = \bigcap_{i\in J} U_i.$$

[Note that, since $J \neq \varnothing$, the scheme $U_J$ is separated, so that the right side of the formula is well defined.]

*Example.* Choose for $X$ the simplest non separated scheme, namely the affine line $\mathbf{A}^1 = \mathrm{Spec}\,\mathbf{Z}[t]$ with the 0-section blown up into two. It is covered by two copies of $\mathbf{A}^1$, with intersection $B = \mathrm{Spec}\,\mathbf{Z}[t, t^{-1}]$. By following the above recipe, we find

$$h_{X,\ell} \;=\; h_{\mathbf{A}^1,\ell} + h_{\mathbf{A}^1,\ell} - h_{B,\ell} \;=\; 2\chi_\ell^{-1} - (\chi_\ell^{-1} - 1) \;=\; \chi_\ell^{-1} + 1,$$

which is the same $\ell$-adic character as for the projective line. This fits with the fact that $N_X(p) = p + 1$ for every $p$.

### 6.1.2. Application : the frobenian property of $N_X(p)$ mod $m$

In what follows, for each prime $\ell$, $S_\ell$ denotes a finite set of primes that is large enough for Theorem 6.1 to apply.

**Theorem 6.3.** *Let $m$ be an integer $\geqslant 1$, and let $S_m$ be the union of the $S_\ell$ for $\ell | m$.*

a) *The function $p \mapsto N_X(p) \pmod{m}$ from $P - S_m$ to $\mathbf{Z}/m\mathbf{Z}$ is $S_m$-frobenian in the sense of* §3.3.1.

b) *The value at* 1 *of that function* (in the sense of §3.3.2.2) *is the image in $\mathbf{Z}/m\mathbf{Z}$ of the Euler-Poincaré characteristic $\chi(X(\mathbf{C}))$ of $X(\mathbf{C})$.*

c) *The value at* $-1$ *of that function* (in the sense of §3.3.2.2) *is the image in $\mathbf{Z}/m\mathbf{Z}$ of the Euler-Poincaré characteristic with compact support $\chi_c(X(\mathbf{R}))$ of $X(\mathbf{R})$.*

d) *If $e \geqslant 1$, the $\Psi^e$-transform* (in the sense of §3.3.2.3) *of that function is $p \mapsto N_X(p^e) \pmod{m}$.*

[Part d) is in fact true for every $e \in \mathbf{Z}$; note that, if $e \leqslant 0$, $N_X(p^e)$ belongs to $\mathbf{Z}[1/p]$ and, since $p \nmid m$ , its reduction mod $m$ is well defined.]

*Note.* When $X$ is not separated, the Euler-Poincaré characteristics of b) and c) are defined by additivity, by the same method as in the last section : choose an open finite covering $(U_i)_{i \in I}$ of $X$, where the $U_i$ are separated, and define $\chi(X(\mathbf{C}))$ by the formula :

$$\chi(X(\mathbf{C})) = \sum_{J \subset I, J \neq \varnothing} (-1)^{|J|+1} \chi(U_J(\mathbf{C})), \quad \text{where } U_J = \bigcap_{i \in J} U_i.$$

The definition of $\chi_c(X(\mathbf{R}))$ is similar.

This shows that it is enough to prove Theorem 6.3 when $X$ is separated. The same remark applies to most of the theorems below.

*Proof of Theorem* 6.3. We may assume that $X$ is separated, and also that $m$ is a power $\ell^n$ of a prime number $\ell$.

For every $i$, let us choose a lattice[1] $L_i$ of $H^i(X, \ell)$ that is stable under the action of $\Gamma_{S_\ell}$ ; a possible choice is to take for $L_i$ the image of $H^i_c(\overline{X}, \mathbf{Z}_\ell)$ in $H^i_c(\overline{X}, \mathbf{Q}_\ell) = H^i(X, \ell)$. The group $L_i/\ell^n L_i$ is a free $\mathbf{Z}/\ell^n\mathbf{Z}$-module of rank $B_i = \dim H^i(X, \ell)$. The group $G = \Gamma_{S_\ell}$ acts on each $L_i/\ell^n L_i$ ; by Theorem 6.1, we have

$$N_X(p) \equiv \sum_i (-1)^i \mathrm{Tr}(g_p | L_i/\ell^n L_i) \bmod \ell^n \quad \text{if } p \notin S_\ell.$$

[1] A lattice in a $\mathbf{Q}_\ell$-vector space $V$ of finite dimension $n$ is a free $\mathbf{Z}_\ell$-submodule $L$ of $V$ of rank $n$; the natural map $\mathbf{Q}_\ell \otimes_{\mathbf{Z}_\ell} L \to V$ is then an isomorphism.

This shows that $N_X(p) \bmod \ell^n$ depends only on the image of $g_p$ in the finite group $\prod \mathbf{GL}_{B_i}(\mathbf{Z}/\ell^n\mathbf{Z})$; this proves a). With the notation of §3.3.2, the corresponding map $\varphi : G \to \mathbf{Z}/\ell^n\mathbf{Z}$ is $g \mapsto \sum_i (-1)^i \mathrm{Tr}(g^{-1}|L_i/\ell^n L_i)$. Similarly, if $e \geqslant 1$, we have

$$N_X(p^e) \equiv \sum_i (-1)^i \mathrm{Tr}(g^e_{p,i,n}) = \varphi(\sigma_p^e) \bmod \ell^n \text{ if } p \notin S_\ell,$$

which proves d).

As for b), it follows from the formula defining the "value at 1" , namely

$$\varphi(1) = \sum_i (-1)^i \mathrm{Tr}(1|L_i/\ell^n L_i) = \sum_i (-1)^i B_i \bmod \ell^n,$$

since $\chi(X(\mathbf{C})) = \sum_i (-1)^i B_i$ by Artin's comparison theorem, cf. §.4.2. Similarly, the value at $-1$ of $p \mapsto N_X(p) \bmod \ell^n$ is :

$$\varphi(-1) = \sum_i (-1)^i \mathrm{Tr}(c|L_i/\ell^n L_i),$$

where $c$ is the complex conjugation. If we denote by $c_i$ the trace of $c$ acting on $H^i(X(\mathbf{C}), \mathbf{Q})$, it follows from Artin's theorem that

$$\varphi(-1) = \sum_i (-1)^i c_i \bmod \ell^n.$$

Assertion c) then follows from the following topological formula :

$$\sum_i (-1)^i c_i = \chi_c(X(\mathbf{R})).$$

[If $\sigma$ is an involution of a "reasonable"[2] space $T$, and if $T^\sigma$ denotes the set of fixed points of $\sigma$, one may define the $c_i(T)$ as above, and one has

$$\sum_i (-1)^i c_i(T) = \chi_c(T^\sigma).$$

Indeed, by additivity, this is equivalent to

$$\sum_i (-1)^i c_i(T - T^\sigma) = 0,$$

which is true because $\sigma$ acts freely on $T - T^\sigma$.]

[2] One needs some finiteness properties of $T$ and $T^\sigma$ ; these properties are satisfied here, if only because of the triangulation theorems for real algebraic spaces, cf. [BCR 98, §9.2] ; note that these theorems apply when $X_0$ is affine.

**Corollary 6.4.** *Let $a$ and $m$ be integers with $m \geqslant 1$. The set of primes $p$ such that $N_X(p) \equiv a \pmod m$ has a density, which is a rational number. If $a$ is equal to either $\chi(X(\mathbf{C}))$ or $\chi_c(X(\mathbf{R}))$, that number is $> 0$.*

This follows from the theorem, combined with the Chebotarev density theorem ; see §3.3.2.2.

*Remarks.*

1) Note that, in part c) of Theorem 6.2, it is essential to use $\chi_c$ and not $\chi$, since $\chi(X(\mathbf{R}))$ and $\chi_c(X(\mathbf{R}))$ are usually distinct, cf. §1.4, Remark.

2) It is tempting to use the notation "$N_X(1)$" and "$N_X(-1)$", as if 1 and $-1$ were primes, so as to be able to state b) and c) above in the following suggestive form :

**Formulae 6.5.** *We have :*

$$N_X(1) = \chi_c(X(\mathbf{C})) = \chi(X(\mathbf{C})) \quad \textit{and} \quad N_X(-1) = \chi_c(X(\mathbf{R})).$$

See exercise 2 below for a case where this notation works very well. Note also that, if $S$ is chosen as in Theorem 6.1, we have $N_X(1) = N_X(p^0)$ for every $p \notin S$.

*Exercises.*

1) Let $X$ be the affine plane curve with equation $x^2 + y^2 = 0$, as in §1.3, and let $m$ be an odd positive integer. Show that $\chi(X(\mathbf{C})) = 1$ and that the density of the $p$ with $N_X(p) \equiv 1 \pmod m$ is equal to $\frac{1}{2} + \frac{1}{2\varphi(m)}$.

2) Suppose that, for $p$ large enough, we have

$$N_X(p) = \sum_i m_i \psi_i(p) p^{n_i},$$

where the $m_i, n_i$ are integers with $n_i \geqslant 0$ and the $\psi_i$ are Dirichlet characters. Show :

a) $N_X(p^e) = \sum m_i \psi_i(p)^e p^{en_i}$, for $e \geqslant 1$ and $p$ large enough,

b) $\chi(X(\mathbf{C})) = \sum m_i$ and $\chi_c(X(\mathbf{R})) = \sum (-1)^{n_i} m_i \psi_i(-1)$.

[Hint. Translate the formula giving $N_X(p)$ into an explicit description of the virtual character $h_{X,\ell}$, and apply Formulae 6.5.]

For instance, the projective plane $\mathbf{P}_2$ has $p^2+p+1$ points in $\mathbf{F}_p$ ; by replacing $p$ by 1, we see that the Euler-Poincaré characteristic of $\mathbf{P}_2(\mathbf{C})$ is $1+1+1=3$ ; similarly, the Euler-Poincaré characteristic of $\mathbf{P}_2(\mathbf{R})$ is $1-1+1=1$.

3) Suppose that, for $p$ large enough, we have

$$N_X(p) = \sum m_i . p^{n_i} + a_p,$$

where the $m_i, n_i$ are integers with $n_i \geqslant 0$, and $a_p$ is the $p$-th coefficient of a modular form $\sum a_n q^n$ of positive integral weight on some congruence subgroup $\Gamma_0(N)$ of $\mathbf{SL}_2(\mathbf{Z})$. Show that $\chi(X(\mathbf{C})) = 2a_1 + \sum m_i$, and $\chi_c(X(\mathbf{R})) = \sum(-1)^{n_i} m_i$. Check these formulae in the special case where $X_0$ is an elliptic curve.

[Hint. Same method as for Exercise 1, combined with the existence of odd Galois representations associated with modular forms.]

4) Suppose that $X_0$ is a smooth projective variety, whose connected components have odd dimension. Let $m$ be an integer $\geqslant 1$. Show :

a) There are infinitely many prime $p$ such that $N_X(p) \equiv 0 \pmod 2$ and $p \equiv 1 \pmod m$.

b) There are infinitely many prime $p$ such that $N_X(p) \equiv 0 \pmod 2$ and $p \equiv -1 \pmod m$.

[Hint. Use the Hodge decomposition of $H^\bullet(X(\mathbf{C}), \mathbf{C})$ to show that $\chi(X(\mathbf{C}))$ and $\chi_c(X(\mathbf{R}))$ are even, and define a frobenian function of $p$ with values in $\mathbf{Z}/2\mathbf{Z} \times (\mathbf{Z}/m\mathbf{Z})^\times$ whose value at 1 is $(0,1)$ and at $-1$ is $(0,-1)$.]

### 6.1.3. Application : the relation $N_X(p) = N_Y(p)$

Let us now prove the "rigidity theorem" stated as Theorem 1.3 in Chapter 1, namely :

**Theorem 6.6.** *Let $X, Y$ be two schemes of finite type over $\mathbf{Z}$. Assume that $N_X(p) = N_Y(p)$ for a set of primes of density 1. Then there exists a prime number $p_0$ such that $N_X(p^e) = N_Y(p^e)$ for all $p \geqslant p_0$ and all $e$.*

*Proof.* Choose a prime $\ell$ together with a finite set of primes $S_\ell$ that is large enough so that Theorem 6.1 applies to both $X$ and $Y$. Let $m$ be a power of $\ell$. By Theorem 6.2 a), the two functions $p \mapsto N_X(p) \pmod m$ and $p \mapsto N_X(p) \pmod m$ are $S_\ell$-frobenian. Since they coincide on a set of $p$ of density 1, they coincide, cf. §3.3.2.1. By Theorem 6.2 d), their $\Psi^e$-transforms are respectively $p \mapsto N_X(p^e) \pmod m$ and $p \mapsto N_Y(p^e) \pmod m$. This shows that $N_X(p^e) \equiv N_Y(p^e) \pmod m$ for every $p \notin S_\ell$. Since this is true for every power $m$ of $\ell$, we have $N_X(p^e) = N_Y(p^e)$ for every $p \in P - S_\ell$, as asserted.

*Remark.* We shall see in §6.3.3 (cf. Theorem 6.15) that the "density 1" condition of Theorem 6.6 can be relaxed to " density $> 1 - 1/B^2$ ", where $B = \sum B_i(X) + \sum B_i(Y)$.

*Examples of schemes with the same number of points mod p.* Let us assume that $X$ and $Y$ are separated, so that their $\ell$-adic cohomology groups $H^i(X, \ell)$ and $H^i(Y, \ell)$ are defined. If these groups are *isomorphic as Galois modules* over some $\Gamma_S$, it follows from Theorem 6.1 that $X$ and $Y$ have the same number of points in all the $\mathbf{F}_{p^e}$, for $p$ large enough. This happens for instance in the following cases :

1) $X_0$ and $Y_0$[3] are $\mathbf{Q}$-isogenous abelian varieties, or $\mathbf{Q}$-isogenous connected reductive groups.

---

[3] We use for $Y$ the same convention as for $X$, namely we denote by $Y_0$ the corresponding $\mathbf{Q}$-variety.

2) The variety $Y_0$ is a *G-twist* of $X_0$, where $G$ is a connected algebraic group acting on $X_0$. For instance, $X_0$ and $Y_0$ are smooth quadrics in $\mathbf{P}_3$ defined by quadratic forms with the same discriminant.

3) $Y = X/G$, where $G$ is a finite group acting on $X$ in such a way that the quotient $X/G$ exists (this is always the case if $X$ is quasi-projective), and that the action of $G$ on the $H^i(X, \ell)$ is trivial [note that this condition is independent of the choice of $\ell$, because of Artin's comparison theorem]. Indeed, for each $i$, it is known that the natural map $H^i(X/G, \ell) \to H^i(X, \ell)$ is injective and that its image is the subspace of $H^i(X, \ell)$ fixed under the action of $G$, i.e. $H^i(X, \ell)$ itself.

This applies[4] when $X$ is the affine $n$-space $\mathbf{A}^n$ since its only non-zero cohomology group is $H^{2n}(X, \ell)$, which is canonically isomorphic to $\mathbf{Q}_\ell(-n)$, hence is fixed under the action of $G$. (We thus get "affine-looking schemes"; we shall give a characterization of them in §7.2.5.) The same is true when $\mathbf{A}^n$ is replaced by $\mathbf{P}_n$.

For instance, in characteristic $\neq 2$ the quadratic cone in $\mathbf{A}^3$ defined by the equation $x^2 = yz$ is isomorphic to the quotient of the affine plane $\mathbf{A}^2$ by the group $G = \{\pm 1\}$. This "explains"[5] that the number of $\mathbf{F}_q$-points of that cone is $q^2$, cf. §2.3.1. (This also works in characteristic 2, at the cost of replacing $\{\pm 1\}$ by the group scheme $\mu_2$.)

*Exercise.* Assume that $X_0$ and $Y_0$ are reduced and of dimension 0. Choose a finite Galois extension $E$ of $\mathbf{Q}$, with Galois group $G$, such that $\mathrm{Gal}(\overline{\mathbf{Q}}/E)$ acts trivially on the finite sets $\Omega_X = X(\overline{\mathbf{Q}})$ and $\Omega_Y = Y(\overline{\mathbf{Q}})$.

a) The schemes $X_0$ and $Y_0$ are isomorphic if and only if the $G$-sets $\Omega_X$ and $\Omega_Y$ are isomorphic.

b) Show the equivalence of the following properties :

$b_1$) The hypothesis of Theorem 6.6 is satisfied, i.e. $N_X(p) = N_Y(p)$ for almost all $p$.

$b_2$) For every $g \in G$, the number of fixed points of $g$ in $\Omega_X$ is the same as in $\Omega_Y$.

$b_3$) The Galois modules $H^0(X, \ell)$ and $H^0(Y, \ell)$ are isomorphic for at least one prime $\ell$ (and hence for every prime $\ell$) [6].

$b_4$) $|\Omega_X/H| = |\Omega_Y/H|$ for every subgroup $H$ of $G$[7].

---

[4]This example was pointed out to me by G. Lusztig.

[5]A more down-to-earth explanation is that both the cone and $\mathbf{A}^2$ contain a copy of $\mathbf{A}^1$ whose complement is isomorphic to $\mathbf{A}^1 \times (\mathbf{A}^1 - \{0\})$.

[6]Motivic interpretation : the $\mathbf{Q}$-motives defined by $X$ and $Y$ are isomorphic.

[7]In the terminology introduced in exerc. 5 of [Se 78, §13.1], properties $b_2$), $b_3$) and $b_4$) mean that the $G$-sets $\Omega_X$ and $\Omega_Y$ are *weakly isomorphic*. The reader should be warned that there was a mistake in the statement of that exercise in the French 1971 edition of [Se 78].

c) Assume $b_1)$, ... , $b_4)$ and let $n = |\Omega_X| = |\Omega_Y|$. Show that, if $n \leqslant 5$, condition a) is satisfied, i.e. $X_0$ and $Y_0$ are isomorphic. Show that this does not extend to $n = 6$.

[Hint. Take for $G$ an abelian group of type (2,2), and make it act on a set with 6 elements with orbits of size (1,1,4) and also with orbits of size (2,2,2).]

d) Assume further that $X_0$ and $Y_0$ are irreducible, i.e. that $n > 0$ and that the actions of $G$ on $\Omega_X$ and on $\Omega_Y$ are transitive. This is equivalent to saying that $\Omega_X$ and $\Omega_Y$ are isomorphic to $\operatorname{Spec} K$ and $\operatorname{Spec} L$, where $K$ and $L$ are finite extensions of $\mathbf{Q}$ of degree $n$. Suppose $b_1)$, ..., $b_4)$ hold (the fields $K$ and $L$ are then called *arithmetically equivalent*). Give an example, with $n = 7$, for which $K$ and $L$ are not isomorphic.

[Hint. Choose a Galois extension $E$ of $\mathbf{Q}$ with Galois group $G = \mathbf{SL}_3(\mathbf{F}_2)$, cf. [La 80]; the group $G$ is the automorphism group of the projective plane $\mathbf{P}_2$ over $\mathbf{F}_2$; use the action of $G$ on the set of rational points, and on the set of rational lines; both sets have 7 elements.]

## 6.2. Density properties

### 6.2.1. Chebotarev theorem for infinite extensions

Let $E$ be a Galois extension of $\mathbf{Q}$ (possibly infinite) that is unramified outside a finite set of primes $S$; let $G$ be its Galois group. If $p \notin S$, let $\sigma_p \in G$ denote its (arithmetic) Frobenius element, defined by projective limit from the case where $G$ is finite; as usual, it is only defined up to conjugation. Let $g_p$ be the geometric Frobenius, i.e. $\sigma_p^{-1}$. The $g_p$ (and also the $\sigma_p$) are dense in $G$. More precisely :

**Theorem 6.7.** *Let $Q$ be a subset of $P - S$ of density 1. Then the $\sigma_p$ $(p \in Q)$ are dense in $\operatorname{Cl} G$; the same is true for the $g_p$.*

[Equivalently : the only open subset $U$ of $G$ that is stable under conjugation and contains all the $g_p$, for $p \in Q$, is $G$ itself.]

*Proof.* This follows by a projective limit argument from the case where $G$ is finite, in which case it is a consequence of the Chebotarev density theorem, cf. §3.2.2.

*Remark.* The hypothesis $\operatorname{dens}(Q) = 1$ can be weakened to $\operatorname{upper-dens}(Q) = 1$.

Let now $C$ be a subset of $G$ that is stable under conjugation and let $P_C$ be the set of all $p \notin S$ such that $g_p \in C$. Let us compare the Haar measure $\mu(C)$ of $C$ with the density of $P_C$ :

**Theorem 6.8.**
a) *If* $C$ *is closed, then* upper-dens$(P_C) \leqslant \mu(C)$.
b) *If* $C$ *is open, then* lower-dens$(P_C) \geqslant \mu(C)$.
c) *If* $C$ *is open and closed*[8], *then* $P_C$ *is* $S$*-frobenian* (*cf.* §3.3.1) *of density* $\mu(C)$.

*Proof.* Let us begin with case c). Since $C$ is open and closed, there exists an open normal subgroup $U$ of $G$ such that $C$ is a union of $U$-cosets, so that one may replace $G$ by the finite group $G/U$. It is then clear that $P_C$ is $S$-frobenian of density $\mu(C)$, cf. §3.3.1.

In case a), $\mu(C) = \inf \mu(C')$, where $C'$ runs through the open and closed subsets of $G$ containing $C$, cf. §5.2.1.1. By c) we have upper-dens$(P_C) \leqslant \mu(C')$ for every $C'$, hence the result.

Case b) follows from case a), applied to $G - C$.

**Corollary 6.9.** *If* $C$ *is closed and its measure is* 0, *then* $P_C$ *has density* 0.

*Proof.* This follows from a), since "density 0" $\Longleftrightarrow$ "upper density 0".

**Corollary 6.10.** *Suppose that* $C$ *is quarrable* (cf. [INT, Chap.IV, §5, exerc. 17 d]), *i.e. that the measure of its boundary is* 0. *Then* $P_C$ *has a density, and that density is equal to* $\mu(C)$.

*Proof.* Let $F$ (resp. $O$) be the closure (resp. the interior) of $C$. The boundary $B$ of $C$ is $F - O$. The hypothesis $\mu(B) = 0$ is equivalent to $\mu(F) = \mu(O)$; since $\mu(O) \leqslant \mu(C) \leqslant \mu(F)$ this shows that $\mu(C) = \mu(F) = \mu(O)$. By Theorem 6.7, we have :

$$\text{upper-dens}(P_C) \leqslant \text{upper-dens}(P_F) \leqslant \mu(F) = \mu(C),$$

and

$$\text{lower-dens}(P_C) \geqslant \text{lower-dens}(P_O) \geqslant \mu(O) = \mu(C).$$

This shows that $P_C$ has density $\mu(C)$.

*Exercises*

1) In Theorem 6.7, replace the hypothesis on the subset $Q$ by " $Q$ is frobenian ". Show that there exists a finite subset $F$ of $Q$ such that the closure of the $\sigma_p$, $p \in Q - F$, is open in $\mathrm{Cl}\, G$.

2) Let $Q$ be the set of primes $p$ such that $v_3(1-p)$ is odd, where $v_3$ is the 3-adic valuation of $\mathbf{Q}$.

a) Show that $Q$ has density $3/8$ .

[Hint. Take for $E$ the field generated by the $3^m$-th roots of unity; the corresponding Galois group $G$ can be identified with $\mathbf{Z}_3^\times$. Let $C$ be the open subset of $\mathbf{Z}_3^\times$

[8] Topologists often replace "open and closed" by the portmanteau adjective *clopen*.

made up of the elements $u \neq 1$ such that $v_3(1-u)$ is odd. The boundary of $C$ is $\{1\}$, hence has measure 0. Apply Corollary 6.10 to $Q = P_C$.]

b) Show that $Q$ is not frobenian.

[Hint. Use Exercise 1.]

### 6.2.2. The $\ell$-adic case

We keep the notation and hypotheses of the above §, and we assume that *the Galois group $G$ is an $\ell$-adic Lie group* (i.e. a $\mathbf{Q}_\ell$-analytic group), for instance[9] a compact subgroup of $\mathbf{GL}_n(\mathbf{Q}_\ell)$ for some $n$. Assume also that the subset $C$ of $G$ is a closed analytic subspace of $G$.

**Theorem 6.11.** *The set $P_C$ is the disjoint union of an $S$-frobenian set $P_1$ and a set $P_0$ of density* 0; *we have* $\mathrm{dens}(P_C) = 0$ *if and only if the interior of $C$ is empty.*

[Recall that $P_C$ is the set of all $p \notin S$ such that $\sigma_p \in C$.]

*Proof.* By Proposition 5.9, we have $C = C_1 \sqcup C_0$, where $C_1$ is the interior of $C$ (which is closed because $C$ is analytic) and $C_0$ is closed with empty interior, and has measure 0 . If we define $P_1$ as $P_{C_1}$ and $P_0$ as $P_{C_0}$, we get a splitting of $P_C$ as $P_C = P_1 \sqcup P_0$; this splitting has the required properties : it is clear that $P_1$ is $S$-frobenian, and $P_0$ has density 0 because of Corollary 6.9. As for $P_1$, its density is equal to the measure of $C$; it is 0 (in which case $P_1$ is empty) if and only if the interior of $C$ is empty, cf. Corollary 5.10.

*Remark.* If $x$ is a real number, let us denote by $\pi_0(x)$ the number of $p \in P_0$ with $p \leqslant x$. Since dens $P_0 = 0$, we have $\pi_0(x) = o(x/\log x)$ for $x \to \infty$. This estimate can be sharpened by using [Se 81, Th.10] : one has

$$\pi_0(x) = O(x/(\log x)^{1+\delta}) \text{ for some } \delta > 0,$$

and under GRH :

$$\pi_0(x) = O(x^{1-\delta}) \text{ for some } \delta > 0.$$

It seems likely that $\pi_0(x)$ is at most $O(x^{\frac{1}{2}})$, and maybe even $O(x^{\frac{1}{2}}/\log x)$.

**Corollary 6.12.** *Let us write the "supernatural order" $o_G$ of the $\ell$-adic group $G$ as $o_G = \ell^n.d_G$, with $n \in \{0, 1, ..., \infty\}$ and $d_G \in \mathbf{N}$, with $(\ell, d_G) = 1$. Then the denominator of* $\mathrm{dens}(P_C)$ *is of the form $\ell^{n'}.d$, with $n' \leqslant n$ and $d | d_G$.*

[9] As noticed by Lubotzky, it follows from Ado's theorem on Lie algebras that every compact $\ell$-adic Lie group is embeddable in some $\mathbf{GL}_n(\mathbf{Q}_\ell)$.

[The *supernatural order* of a profinite group $G$ is the l.c.m. of $(G : U)$, when $U$ runs through the open subgroups of $G$; see [Se 64, I.1.3], or [RZ 00, 2.3].]

*Proof.* Indeed, the Haar measure of an open and closed subset of $G$ has a denominator that divides $o_G$.

This corollary is especially useful when $G$ is a subgroup of $\mathbf{GL}_n(\mathbf{Q}_\ell)$, since in that case $o_G$ divides $o_{\mathbf{GL}_n(\mathbf{Z}_\ell)}$, which is equal to $\ell^\infty \prod_{i=1}^n (\ell^i - 1)$; one then sees that the $\ell'$-factor[10] of the denominator of dens$(P_C)$ divides $\prod_{i=1}^n (\ell^i - 1)$.

*Application: relations between $N_X(p)$ and $N_Y(p)$.* Let $X$ and $Y$ be two schemes of finite type over Spec $\mathbf{Z}$, and let us assume that the set $S$ is large enough so that Theorem 6.1 applies to both $(X, \ell, S)$ and $(Y, \ell, S)$ : there are continuous virtual characters $h_X$ and $h_Y$ of $\Gamma_S$ over $\mathbf{Q}_\ell$,

$$N_X(p^e) = h_X(g_p^e) \text{ and } N_Y(p^e) = h_Y(g_p^e) \text{ for every } p \notin S \text{ and every } e \geqslant 1.$$

These characters factor through a quotient $G$ of $\Gamma_S$ which is a subgroup of a product of linear $\ell$-adic groups, hence is an $\ell$-adic group. We may thus apply Theorem 6.11, and we obtain :

**Theorem 6.13.** *Let $F$ be a polynomial in two variables, with coefficients in $\mathbf{Q}$. Let $P_F$ be the set of $p$ such that $F(N_X(p), N_Y(p)) = 0$. Then $P_F$ is the disjoint union of a frobenian set and a set of density* 0.

*Remark.* There is a similar result for equations involving not only $N_X(p)$ and $N_Y(p)$ but also $N_X(p^2)$, etc. We shall look into such an example in §7.2.5.

*Example.* Suppose $X_0$ is an elliptic curve, and let $Y = \mathbf{P}_1$. Choose for $F$ the polynomial $F(u, v) = u - v$, so that $P_F$ is the set of primes $p$ such that $N_X(p) = p + 1$; if we write $N_X(p)$ as $p + 1 - a_p$, cf. §4.7.1.2, this means that $a_p = 0$. There are two cases :

1) The curve $X_0$ has complex multiplication (over $\overline{\mathbf{Q}}$) by an imaginary quadratic field $K$. Let $\varepsilon$ be the quadratic character of $\Gamma_\mathbf{Q}$ associated with $K$. Except for a finite number of $p$, one finds that $a_p = 0 \iff \varepsilon(p) = -1$. In that case, $P_F$ is frobenian of density $\frac{1}{2}$.

2) The curve $X_0$ does not have complex multiplication. In that case $P_F$ has density 0 since the subvariety of $\mathbf{GL}_2$ given by the equation "Trace $= 0$" has dimension $< \dim \mathbf{GL}_2$, cf. Theorem 6.11. By a theorem of Elkies

[10] If $m$ is a non-zero integer, the *$\ell'$-factor* of $m$ is the largest divisor of $m$ that is prime to $\ell$.

([El 87]), $P_F$ is infinite. It is conjectured, cf. [LT 76], that the number of $p \leqslant x$ with $p \in P_F$ is of the order of magnitude of $x^{\frac{1}{2}}/\log x$ when $x \to \infty$; it is known (Elkies-Murty, cf. [El 91]) that it is $\ll x^{3/4}$.

*Exercises.* (The aim of these two exercises is to extend Theorem 6.13 to polynomials with coefficients in an arbitrary commutative ring.)

1) Let $R$ be a commutative ring which is finitely generated over $\mathbf{Z}$, and let $R_t$ be the torsion subgroup of $R$. Choose an integer $m > 0$ such that $mR_t = 0$ (such an $m$ exists because $R$ is noetherian).

a) Show that $R$ is isomorphic, as an additive group, to the direct sum of $R_t$ and a torsion-free group.
[Hint. Show that the group $E$ =Ext$(R/R_t, R_t)$ is 0 by proving that the map $m : E \to E$ is 0 (because $m$ kills $R_t$) and is surjective (because $R/R_t$ is torsion-free). See also [Ka 52, Cor. to Th.5] and [A IV-VII, chap.VII, §2, Exerc.7d].]

b) Show that there exist two families of additive homomorphisms

$$f_\lambda : R \to \mathbf{Q} \quad \text{and} \quad g_\mu : R \to \mathbf{Z}/m\mathbf{Z}$$

that give an injection of $R$ into a direct sum of copies of $\mathbf{Q}$ and of $\mathbf{Z}/m\mathbf{Z}$.
[Hint. Use a) combined with the fact, cf. [Pr 23], that every abelian group of finite exponent is isomorphic to a direct sum of cyclic groups, cf. [A IV-VII, chap.VII, §2, Exerc.4c].]

2) Let $n$ and $m$ be positive integers, with $m \geqslant 1$. Let us say that a subset $E$ of $\mathbf{Z}^n$ is *m-elementary* if there exist a subset $E_m$ of $\mathbf{Z}^n/m\mathbf{Z}^n$ and a polynomial $P \in \mathbf{Q}[X_1, ..., X_n]$ such that $E$ is the set of all $x \in \mathbf{Z}^n$ such that $P(x) = 0$ and $x$ belongs to $E_m$ mod $m$.

a) Show that every intersection of $m$-elementary subsets is $m$-elementary.
[Hint. Note that an infinite number of polynomials can be reduced to a finite number of such, and (by taking a sum of squares) to just one such.]

b) Let $R$ be a commutative ring, and let $Y = Y_1, ..., Y_r$ be a finite family of elements of $R[X_1, ..., X_n]$. Let $E_Y$ be the set of all $x \in \mathbf{Z}^n$ such that $Y_i(x) = 0$ in $R$ for every $i$. Show that $E_Y$ is $m$-elementary for a suitable choice of $m$.
[Hint. One may replace $R$ by the subring generated by the coefficients of the $Y_i$, hence assume that $R$ is finitely generated over $\mathbf{Z}$. Let $m, f_\lambda, g_\mu$ be as in Exerc.1. By applying $f_\lambda$ to the coefficients of $Y_i$, one gets a polynomial $Y_{i,\lambda}$ with coefficients in $\mathbf{Q}$. Similarly, by applying $g_\mu$ to the coefficients of $Y_i$, one gets a polynomial $Y_{i,\mu}$ with coefficients in $\mathbf{Z}/m\mathbf{Z}$. Use the polynomials $Y_{i,\lambda}$ and $Y_{i,\mu}$ to show that $E_Y$ is an intersection of $m$-elementary sets, hence is $m$-elementary by a).]

c) Let $E$ be an $m$-elementary subset of $\mathbf{Z}^2$. Show that the set of $p$ such that $(N_X(p), N_Y(p))$ belongs to $E$ is the disjoint union of a frobenian set and a set of density 0.
[Hint. Same proof as for Theorem 6.13, using Theorem 6.3.]

d) Use b) and c) to show that Theorem 6.13 remains valid for a polynomial with coefficients in an arbitrary commutative ring.

## 6.3. About $N_X(p) - N_Y(p)$

We keep the same notation $(X, Y, \ell, S...)$ as above; we are going to look more closely at the possible values of the function $N_X(p) - N_Y(p)$.

Define a virtual $\ell$-adic character $h$ of $\Gamma_S$ by the formula $h = h_X - h_Y$; we have

$$h(g_p^e) = N_X(p^e) - N_Y(p^e) \text{ if } p \notin S \text{ and } e \in \mathbf{Z}.$$

### 6.3.1. The case where $N_X(p)$ and $N_Y(p)$ are close to each other

**Theorem 6.14.** *Suppose $|N_X(p) - N_Y(p)|$ remains bounded when $p$ varies. Then the function $p \mapsto N_X(p) - N_Y(p)$ is $S$-frobenian.*

*Proof.* Let $I$ be the set of values of $h(g_p) = N_X(p) - N_Y(p)$. By assumption, it is a finite set. Since the $g_p$ are dense in $\mathrm{Cl}(\Gamma_S)$, we have $h(g) \in I$ for every $g \in \Gamma_S$. Hence $h$ is a virtual character that only takes finitely many values. By Theorem 5.5, this implies the existence of a normal subgroup $N$ of $\Gamma_S$, of finite index, such that $h$ is constant on every $N$-coset. Let $\overline{N}$ be the closure of $N$; since $h$ is continuous, it is constant on every $\overline{N}$-coset, i.e. it comes from a class function on the finite Galois group $\Gamma_S/\overline{N}$. Hence the function $p \mapsto h(g_p) = N_X(p) - N_Y(p)$ is an $S$-frobenian map of $P - S$ into $\mathbf{Z}$.

[*Alternative proof.* Since $I$ is finite, there exists an integer $m > 0$ such that the map $I \to \mathbf{Z}/\ell^m\mathbf{Z}$ is injective. Hence it is enough to show that the map $p \mapsto N_X(p) - N_Y(p) \bmod \ell^m$ is $S$-frobenian; but this follows from Theorem 6.2, applied to $X$ and to $Y$.]

*Remarks.*

1) Both proofs also show that $N_X(p^e) - N_Y(p^e)$ belongs to the set $I$ for every $p \notin S$ and every $e \in \mathbf{Z}$.

2) When $Y = \varnothing$, the theorem says that, if $N_X(p)$ remains bounded when $p$ varies, then it is a frobenian function of $p$. This is not surprising : since the $N_X(p^e)$ are also bounded (for $p$ large enough), one sees easily that $\dim X_0 \leqslant 0$, so that we can apply §3.4.2.1.

3) The hypothesis that $|N_X(p) - N_Y(p)|$ is bounded could be weakened to : there exists a subset $Q$ of $P$, of density 1, such that $p \mapsto |N_X(p) - N_Y(p)|$ is bounded on $Q$; this follows from the fact that the $g_p$, for $p \in Q$, are dense in $\mathrm{Cl}\,\Gamma_S$, cf. Theorem 6.7.

4) The virtual character $h : \Gamma_S \to \mathbf{Z}$ defined in the proof above is a difference of two permutation characters (see §7.2.3, exerc. 3).

5) When $|N_X(p) - N_Y(p)|$ is unbounded, the Sato-Tate conjecture (cf. Chap.8) implies that, for every $\varepsilon > 0$, there exist infinitely many $p$ with $|N_X(p) - N_Y(p)| > (2-\varepsilon)p^{\frac{1}{2}}$, cf. §8.4.4, exerc. 3.

### 6.3.2. The case where $N_X(p)$ and $N_Y(p)$ are very close to each other

**Theorem 6.15.** *Suppose $N_X(p) - N_Y(p)$ is equal to $0, 1$ or $-1$ for every $p$ in a set of density 1. Then there are only three possibilities :*

a) $N_X(p) = N_Y(p)$ *for every $p \notin S$.*

b) *There exists an integer $m > 0$ and a Dirichlet character*

$$\varepsilon : (\mathbf{Z}/m\mathbf{Z})^\times \to \{1,-1\},$$

*such that all the prime divisors of $m$ belong to $S$, and $N_X(p) - N_Y(p) = \varepsilon(p)$ for every $p \notin S$.*

c) *Same as* b), *with* $N_X(p) - N_Y(p)$ *replaced by* $N_Y(p) - N_X(p)$.

*Proof.* Same method as for Theorem 6.14 above : the fact that the $g_p$ are dense in Cl $\Gamma_S$ implies that the virtual character $h$ takes values in $\{-1, 0, 1\}$, and Theorem 5.2 then shows that $h$ is either 0, or a continuous homomorphism $\varepsilon : \Gamma_S \to \{1,-1\}$, or the negative of such a homomorphism.

*Remark.* Theorem 6.15 remains valid if the function $p \mapsto N_X(p) - N_Y(p)$ is replaced by any polynomial in $N_X(p), N_X(p^2), ..., N_Y(p), N_Y(p^2), ...$ with integer coefficients : if such a function takes values in $\{-1, 0, 1\}$, its restriction to $P - S$ is either 0, or $\varepsilon$ or $-\varepsilon$. The same remark applies to Theorem 6.13.

Note also that b) implies the existence of a non-zero integer $d$ such that $N_X(p) - N_Y(p) = (\frac{d}{p})$ for every $p \notin S$.

**Corollary 6.16.** *If $N_X(p) - N_Y(p)$ is equal to 0 or 1 for every $p$ in a set of density 1, then either $N_X(p) = N_Y(p)$ for every $p \notin S$ or $N_X(p) = 1 + N_Y(p)$ for every $p \notin S$.*

This follows from Theorem 6.15 since case c) is impossible (choose $p$ such that $\varepsilon(p) = 1$), and case b) is possible only if $\varepsilon = 1$.

*Remarks.*

1) There are analogous results for $N_X(p^e) - N_Y(p^e)$. For instance, in case b) of Theorem 6.15 , we have $N_X(p^e) = N_Y(p^e) + \varepsilon(p^e)$ if $p \notin S$ and $e \in \mathbf{Z}$ : this follows from the equality $h = \varepsilon$.

2) All three cases of Theorem 6.15 can occur, as one sees by taking $X_0$ and $Y_0$ of dimension 0. For instance, if we choose $X = \operatorname{Spec} \mathbf{Z}$ and $Y = \operatorname{Spec} \mathbf{Z}[i]$, we have case c), and we can choose $\ell = 2, S = \{\ell\}$ and $\varepsilon(p) = (-1)^{(p-1)/2}$.

### 6.3.3. The density of the set of $p$ with $N_X(p) = N_Y(p)$

In this section, we assume that $X$ and $Y$ are separated, so that their cohomology with proper support is well defined.

Put $B_i(X) = \dim H^i(X, \ell)$ and $B(X) = \sum_i B_i(X)$; by Artin's comparison theorem, we have $B(X) = \sum_i \dim_{\mathbf{Q}} H^i_c(X(\mathbf{C}), \mathbf{Q})$. Define similarly $B_i(Y)$ and $B(Y)$, and put $B = B(X) + B(Y) = B(X \sqcup Y)$.

**Theorem 6.17.** *Assume that $N_X(p) \neq N_Y(p)$ for at least one $p \notin S$. Then the set of $p$ such that $N_X(p) = N_Y(p)$ has density $\leqslant 1 - 1/B^2$.*

*Proof.* By assumption, the virtual character $h = h_X - h_Y$ is not identically zero. We may write it as $h = \chi_1 - \chi_2$, where $\chi_1$ and $\chi_2$ are the (effective) $\ell$-adic characters of $\Gamma_S$ given by the formulae :

$$\chi_1 = \sum_{i \text{ even}} h_{i,X} + \sum_{i \text{ odd}} h_{i,Y},$$

and

$$\chi_2 = \sum_{i \text{ odd}} h_{i,X} + \sum_{i \text{ even}} h_{i,Y}.$$

Let $a$ and $b$ be the degrees of $\chi_1$ and $\chi_2$. We have $a+b = B(X)+B(Y) = B$. By Theorem 5.15, the set $I$ of the elements $g \in \Gamma_S$ with $h(g) = 0$ has a Haar measure $\mu(I) \leqslant 1 - 1/(a+b)\sup(a, b) \leqslant 1 - 1/B^2$. By what we have seen in §6.1.2, the set of $p$ with $g_p \in I$ has a density equal to $\mu(I)$, hence $\leqslant 1 - 1/B^2$.

*Remark.* The term $1/B^2$ is rarely optimal. If, for instance, $X_0$ and $Y_0$ are elliptic curves that are not $\mathbf{Q}$-isogenous, one has $B = 8$, and the theorem says that the density of the $p$ such that $N_X(p) = N_Y(p)$ is $\leqslant 63/64$. In fact, that density is $\leqslant 3/4$, the $3/4$-case occurring when $X_0$ has complex multiplication, and $Y_0$ is a non-trivial quadratic twist of it.

### 6.3.4. A Minkowski-style bound for the denominator of the density

We keep the hypotheses and notation of the last section. Let $P_I$ be the set of primes $p$ such that $N_X(p) = N_Y(p)$; we have shown that $\text{dens}(P_I) \leqslant 1 - 1/B^2$. We are now going to give an upper bound for the denominator of the rational number $\text{dens}(P_I)$. To state the result, let us introduce some notation :

Let $n$ be an integer $\geqslant 1$, and let $M(n)$ denote the least common multiple of the orders of the finite subgroups of $\mathbf{GL}_n(\mathbf{Q})$. By a theorem of Minkowski

([Mi 87], see also [Se 07, §1]), we have

$$M(n) = \prod_p p^{m(n,p)}, \text{ where } m(n,p) = [\frac{n}{p-1}]+[\frac{n}{p(p-1)}]+[\frac{n}{p^2(p-1)}]+\cdots$$

Define $m'(n,p)$ as being $m(n,p)$ if $p \neq 2$ and $m'(n,p) = m(n,p) + [n/2]$ if $p = 2$. Put

$$M'(n) = \prod_p p^{m'(n,p)} = 2^{[n/2]} M(n).$$

We have for instance :

$$M'(1) = 2, M'(2) = 48, M'(3) = 96, M'(4) = 21640, M'(5) = 43280.$$

Our estimate for the denominator of $\mathrm{dens}(P_I)$ involves the function $M'$ :

**Theorem 6.18.** *The denominator of* $\mathrm{dens}(P_I)$ *is a divisor of* $M'(B)$, *where* $B = B(X) + B(Y)$ *as in* §6.3.3.

[For instance, if $B = 3$, the theorem says that $\mathrm{dens}(P_I)$ belongs to the set $\{0, 1/96, 2/96, ..., 95/96, 1\}$.]

*Proof.* We need :

**Lemma 6.19.** *Let $d$ and $n$ be integers $\geqslant 1$. Suppose that, for every prime $\ell$, the $\ell'$-factor of $d$ divides the product $(\ell-1)(\ell^2-1)\cdots(\ell^n-1)$. Then $d$ divides $M'(n)$.*

*Proof of the Lemma.* Let $p$ be a prime number. We have to show that the $p$-adic valuation $v_p(d)$ of $d$ is $\leqslant m'(n,p)$. To do so, let us choose $\ell$ such that :

a) $l \equiv 3 \pmod 8$ in case $p = 2$,

b) the image of $\ell$ in $(\mathbf{Z}/p^2\mathbf{Z})^\times$ is a generator of that group in case $p > 2$.

By assumption, we have

$$v_p(d) \leqslant v_p((\ell-1)(\ell^2-1)\cdots(\ell^n-1)) = \sum_{i=1}^{n} v_p(\ell^i - 1).$$

Assume first that $p > 2$. Property b) above implies that $v_p(\ell^i - 1) = 0$ if $p-1$ does not divide $i$, and $v_p(\ell^i - 1) = 1 + v_p(i)$ if $p-1$ divides $i$. The sum of the $v_p(\ell^i - 1)$ is then easy to compute, and one finds (cf. e.g. [Se 07, §1.3.3]) that it is equal to $m(n,p) = m'(n,p)$.

The case $p = 2$ is analogous : one has $v_2(\ell^i - 1) = 1$ if $i$ is odd, and $v_2(\ell^i - 1) = 2 + v_2(i)$ if $i$ is even.

*Proof of Theorem 6.18, continued.* Let $d$ be the denominator of $\mathrm{dens}(P_I)$ and let $\ell$ be a prime number. If $S$ is a sufficiently large finite set of primes, then $P_I - S \cap P_I$ is of the form $P_C$ for some analytic (and even algebraic) subspace $C$ of $\mathbf{GL}_B(\mathbf{Q}_\ell)$. The density of $P_C$ and the density of $P_I$ are the same. Hence $d$ is the denominator of dens$(P_C)$, and Corollary 6.12 shows that the $\ell'$-factor of $d$ divides the product $(\ell - 1)\cdots(\ell^B - 1)$. Since this is true for every $\ell$, Lemma 6.19 shows that $d$ divides $M'(B)$.

*Remark.* A similar argument gives the slightly better statement that $d$ divides $\prod_i M'(B_i(X))M'(B_i(Y))$. Note also that the same results hold for any algebraic relation between $N_X(p), N_X(p^2), \ldots, N_Y(p), N_Y(p^2), \ldots$; we thus find a *uniform bound* for the denominator of the density of any set of primes defined in such a way.

*Exercise.* Show that Lemma 6.19 remains valid if " for every prime $\ell$ " is replaced by " for every prime $\ell$ belonging to a set of upper density 1 ".

# CHAPTER 7
# THE ARCHIMEDEAN PROPERTIES OF $N_X(p)$

We keep the notation of Chapter 6. In particular, $X$ is a scheme of finite type over $\mathbf{Z}$; we denote by $X_0$ the corresponding $\mathbf{Q}$-variety, i.e. $X_{/\mathbf{Q}}$. The main difference with Chapter 6 is that, instead of looking at congruence properties of $N_X(p)$, we look at its size.

## 7.1. The weight decomposition of the $\ell$-adic character $h_X$

### 7.1.1. The weight of an $\ell$-adic representation

Let $\ell$ be a prime number and let $\rho : \Gamma_{\mathbf{Q}} \to \mathbf{GL}_n(\mathbf{Q}_\ell)$ be a continuous $\ell$-adic representation of $\Gamma_{\mathbf{Q}} = \mathrm{Gal}(\overline{\mathbf{Q}}/\mathbf{Q})$. Let $w$ be an integer $\geqslant 0$. If $S$ is a finite set of primes, we say that $\rho$ *has weight* $w$ *outside* $S$ if :

a) $\rho$ is unramified outside $S$, i.e. $\rho$ can be factored through the group $\Gamma_S$ of Chapter 6.

b) For every prime $p \notin S$, the eigenvalues of $\rho(g_p)$ are $p$-Weil integers of weight $w$ (cf. §4.5).

Recall that $g_p$ means the geometric Frobenius associated with $p$; thanks to a), its conjugacy class in $\Gamma_S$ is well defined.

When there exists such an $S$ we say that $\rho$ *has weight* $w$. Note that the weight of $\rho$, if it exists, is unique (provided that $n \geqslant 1$).

An $\ell$-adic character $f : \Gamma_{\mathbf{Q}} \to \mathbf{Q}_\ell$ is called *of weight* $w$ *outside* $S$ if it is the character of a representation of weight $w$ outside $S$; same convention for *weight* $w$ without mentioning $S$.

These definitions extend to $\ell$-adic virtual characters in an obvious way. If $f$ is such a character, the following two properties are equivalent :

i) $f$ has weight $w$ outside $S$.

ii) If $f = \sum_\chi n_\chi \chi$, where the $\chi$'s are irreducible and distinct, and the integers $n_\chi$ are $\neq 0$, then all the $\chi$'s have weight $w$ outside $S$.

In that case, the values of $f$ on the $g_p^e$, with $p \notin S$ and $e \geqslant 0$, are algebraic integers; their archimedean size is $\ll p^{ew/2}$ in the following sense : there exists $C > 0$ such that, for every $p \notin S$, every $e \geqslant 0$ and every embedding $\iota : \mathbf{Q}_\ell \to \mathbf{C}$, one has $|\iota(f(g_p^e))| \leqslant Cp^{ew/2}$. More precisely, if $f$ is the difference of two characters of degree $a$ and $b$, one may take for $C$ the sum $a+b$. Of course, one would like to say more on the size of

$|\iota(f(g_p^e))|/p^{ew/2}$, especially when $e=1$; we shall come back to this question in the next chapter.

*Example.* Assume $\ell \in S$ and let $\chi_\ell : \Gamma_S \to \mathbf{Z}_\ell^\times$ be the $\ell$-adic *cyclotomic character* of $\Gamma_S$; it is characterized by $\chi_\ell(\sigma_p) = p$, i.e. $\chi_\ell(g_p) = p^{-1}$. The character $\chi_\ell^{-1}$ has weight 2.

The weight 0 case is easy to describe :

**Theorem 7.1.** *Let $f_1$ and $f_2$ be two continuous $\ell$-adic characters of $\Gamma_S$, and let $f$ be the virtual character $f_1 - f_2$. The following properties are equivalent :*

a) *$f$ is of weight* 0 *outside $S$.*

b) *The elements $f(g_p)$, $p \notin S$, belong to a finite subset of $\mathbf{Q}_\ell$.*

c) *$f$ factors through a finite quotient $\Gamma_S/N$, where $N$ is an open normal subgroup of $\Gamma_S$.*

[Note that, in case c), $f$ defines a virtual character of $\Gamma_S/N$, cf. Theorem 5.6.]

*Proof.*

a) $\Rightarrow$ b) : It is enough to prove this when $f$ is the character of a continuous representation $\rho : \Gamma_{\mathbf{Q}} \to \mathbf{GL}_n(\mathbf{Q}_\ell)$ that is of weight 0 outside $S$. For every $p \notin S$, the eigenvalues of $\rho(g_p)$ in $\overline{\mathbf{Q}}_\ell$ are $p$-Weil integers of weight 0, i.e. roots of unity. This means that the semisimple component $\rho(g_p)^{\mathrm{ss}}$ of $\rho(g_p)$ has finite order. But the order of the torsion elements of $\mathbf{GL}_n(\mathbf{Q}_\ell)$ is bounded, a rough bound being for instance $\ell^n \prod_{i=1}^n (\ell^i - 1)$. Hence the trace of $\rho(g_p)^{\mathrm{ss}}$, which is the same as the trace of $\rho(g_p)$, belongs to a finite subset of $\mathbf{Q}_\ell$.

b) $\Rightarrow$ c) : By Theorem 5.5, there exists a normal subgroup $N_0$ of $\Gamma_S$, of finite index, such that $f$ is constant on every $N_0$-coset of $\Gamma_S$. The closure $N = \overline{N}_0$ of $N_0$ has finite index, hence is open; since $f$ is continuous, $f$ is constant on every $N$-coset of $\Gamma_S$.

c) $\Rightarrow$ a) : Clear.

*Weight decompositions.*

If $f$ is a virtual $\ell$-adic character of $\Gamma_{\mathbf{Q}}$, and if $f = \sum_w f_w$, where each $f_w$ has weight $w$, we say that $f$ has a *weight decomposition*; if such a decomposition exists, it is unique : this follows from the fact that irreducible characters with different weights are distinct.

*Basic properties of the weight decompositions.*

We have the standard properties of a grading; for instance, if $f$ has weight $w$ and $f'$ has weight $w'$, then $ff'$ has weight $w + w'$.

The same applies to the $\lambda$ and $\Psi$ operations on virtual characters (cf. §5.1.1.2) : if $f$ has weight $w$ and if $n \in \mathbf{N}$, then the virtual characters $\lambda^n f$ and $\Psi^n f$ have weight $nw$.

[*Proof.* Assume first that $f$ is effective, hence is the character of a representation $\rho$ of weight $w$. In that case, if the eigenvalues of $\rho(g_p)$ are $p$-Weil integers of weight $w$, those of $\wedge^n \rho(g_p)$ are products of $n$ such, which shows that $\wedge^n \rho$ has weight $nw$. The general case follows, because, if $f = g - h$, then both $\lambda^n f$ and $\Psi^n f$ can be written as homogeneous polynomials of weight $n$ in the $\lambda^i g$ and the $\lambda^j h$.]

*Conjugate characters.*

Let $f$ be a virtual character of weight $w$ outside $S$, and assume that $\ell \in S$; let us denote by $\chi_\ell$ the $\ell$-cyclotomic character (see above). The character

$$\chi_\ell^{-w}\Psi^{-1}f, \quad \text{i.e.} \quad \gamma \mapsto \chi_\ell(\gamma)^{-w} f(\gamma^{-1}),$$

has also weight $w$. This character deserves to be denoted by $\overline{f}$ : its values on the powers of the Frobenius elements are the complex conjugates of those of $f$[1]. In particular, the following three properties are equivalent :

$f = \overline{f}$;

$f(g_p)$ is a totally real algebraic integer for every $p \notin S$;

$f(g_p^e)$ is a totally real algebraic number for every $p \notin S$ and every $e \in \mathbf{Z}$.

*Exercise.* Let $(f, S, w)$ be as above.

a) Assume $p \notin S$. Show that $f(g_p^{-e}) = p^{-ew}\overline{f}(g_p^e)$ for every $e \in \mathbf{Z}$.

[Hint. Use the fact that, if $\lambda$ is a $q$-Weil integer of weight $w$, one has $\lambda^{-1} = q^{-w}\bar{\lambda}$.]

b) Assume $w$ is odd and that all the $f(g_p)$ belong to $\mathbf{Z}$. Let $c$ be the complex conjugation in $\Gamma_S$ (it is well defined up to conjugation). Show that $f(c) = 0$.

[Hint. Use the fact that $\chi_\ell(c) = -1$.]

### 7.1.2. Weight decomposition of $h_X$

Recall that $X$ is a scheme of finite type over $\mathbf{Z}$, and $X_0 = X_{/\mathbf{Q}}$. If $\ell$ is a prime number, we have defined in §6.1.1 an $\ell$-adic virtual character of $\Gamma_\mathbf{Q}$ that we denoted by $h_X$; here, we shall write it as $h_{X,\ell}$ when its dependence from $\ell$ will be important. Note that $h_{X,\ell}$ depends only on the $\mathbf{Q}$-variety $X_0$ and on $\ell$; when $X_0$ is separated, it is the alternating sum of the characters $h_{i,X,\ell}$ associated with the Galois representations that we denoted by $H^i(X, \ell)$.

[1]More correctly, for every embedding $\iota : \mathbf{Q}_\ell \to \mathbf{C}$, we have $\iota(\overline{f}(\gamma)) = \overline{\iota(f(\gamma))}$ for every element $\gamma \in \Gamma_S$ that is a power of a $\sigma_p$ with $p \notin S$.

**Theorem 7.2.** *Assume that $X_0$ is separated. Each $h_{i,X,\ell}$ has a weight decomposition in which the weights are $\leqslant i$.*

*Proof.* Let us decompose $h_{i,X,\ell}$ into a sum of irreducible characters, and let $f$ be one of them. We need to prove that $f$ has a weight, and that this weight is $\leqslant i$. This follows from the following two results :

**Proposition 7.3.** *There exists a smooth projective variety $Y$ over $\mathbf{Q}$, of dimension $\leqslant \dim X_0$, and an integer $j \leqslant i$ such that $f$ occurs as an irreducible component of $h_{j,Y,\ell}$.*

*Proof.* We use induction on $\dim X_0$. The case $\dim X_0 = 0$ is trivial. If $\dim X_0 > 0$, we may assume that $X_0$ is reduced, and, by a well-known theorem of Hironaka (resolution of singularities, cf. [Hi 64] and also [Ko 07]), there exists a dense open affine subscheme $U$ of $X_0$ that can be embedded as a dense open subscheme into a smooth projective $\mathbf{Q}$-variety $Z$ ; note that the dimensions of $X_0$, $U$ and $Z$ are the same. We then use the two exact sequences of $\ell$-adic representations :

$$H^i(U,\ell) \to H^i(X_0,\ell) \to H^i(X_0 - U,\ell)$$

and

$$H^{i-1}(Z - U,\ell) \to H^i(U,\ell) \to H^i(Z,\ell).$$

[Recall that $H^i(U,\ell)$ is an abbreviation for $H^i_c(U_{/\overline{\mathbf{Q}}}, \mathbf{Q}_\ell)$, cf. §6.1.1.]

The first exact sequence shows that $f$ occurs either in $h_{i,X_0-U,\ell}$ or in $h_{i,U,\ell}$. In the first case, we apply the induction hypothesis to $X_0 - U$, whose dimension is $< \dim X_0$. In the second case, the second exact sequence shows that, either $f$ occurs in $h_{i,Z,\ell}$, in which case we win since $Z$ is projective and smooth, or $f$ occurs in $h_{i-1,Z-U,\ell}$, and we apply the induction assumption.

*Remark.* By taking hyperplane sections, we could choose $Y$ such that $\dim Y \leqslant i$.

**Proposition 7.4.** *If $X_0$ is proper and smooth over $\mathbf{Q}$, then $h_{i,X,\ell}$ is of weight $i$.*

*Proof.* The hypothesis is equivalent to saying that $X$ is proper and smooth over a dense open subset of Spec $\mathbf{Z}$. If $S$ is the complement of that set, $S$ is a finite set of primes. We may then apply Theorem 4.13 with $S_\ell = S \cup \{\ell\}$ (see the references given in §4.8.4) : if $p \notin S_\ell$, the eigenvalues of $g_p$ on $H^i(X,\ell)$ are the same as those of the Frobenius endomorphism $F$ of $H^i_c(\overline{X_p}, \mathbf{Q}_\ell)$, where $X_p$ is the reduction mod $p$ of $X$. By Deligne's Theorem 4.5, these values are $p$-Weil integers of weight $i$.

*End of the proof of Theorem 7.2.* To prove that $f$ has a weight $\leqslant i$, Proposition 7.3 allows us to assume that $X_0$ is smooth over $\mathbf{Q}$, in which case we apply Proposition 7.4.

**Theorem 7.5.** *The virtual character $h_{X,\ell}$ has a weight decomposition.*

*Proof.* When $X$ is separated, this is a consequence of Theorem 7.2, since $h_{X,\ell}$ is the alternating sum of the $h_{i,X,\ell}$. The general case follows, since, by definition (cf. end of §6.1.1), $h_{X,\ell}$ is a $\mathbf{Z}$-linear combination of some $h_{U,\ell}$, with $U$ separated.

*Notation.* If $i$ is an integer, we shall denote by $h^i_{X,\ell}$ the component of weight $i$ of $h_{X,\ell}$, multiplied by $(-1)^i$. By definition, we have[2]

**Formula 7.6.** $h_{X,\ell} = \sum_i (-1)^i h^i_{X,\ell}$ .

When $X_0$ is proper and smooth, this decomposition coincides with the one used in §6.1.1, i.e. $h^i_{X,\ell} = h_{i,X,\ell}$. (It is in order to have this simple formula that we put the factor $(-1)^i$ in the definition of $h^i_{X,\ell}$.) In general, the characters $h^i_{X,\ell}$ are not effective (except when $i \geqslant 2\dim X_0 - 1$, see §7.1.4 and §7.1.5). A simple example is $X_0 = \mathbf{G}_m = \mathbf{P}_1 - \{0, \infty\}$, where $d = 1, h^2_{X,\ell} = \chi_\ell^{-1}, h^1_{X,\ell} = 0$ and $h^0_{X,\ell} = -1$.

*Exercise.* Show that $\Psi^{-1} h^i_{X,\ell} = \chi_\ell^i h^i_{X,\ell}$. Assume that $X_0$ is projective, smooth, and that all its components have the same dimension $d$. Use Poincaré duality to show that $h^i_{X,\ell} = \chi_\ell^{d-i} h^{2d-i}_{X,\ell}$, and deduce that $\Psi^{-1} h_{X,\ell} = \chi_\ell^d h_{X,\ell}$.

### 7.1.3. Basic properties of the $h^i_{X,\ell}$

The first one is that $h^i_{X,\ell}$ depends *additively* on $X_0$ : if $U \subset X_0$ is open, we have

$$h^i_{X,\ell} = h^i_{U,\ell} + h^i_{X-U,\ell}.$$

This follows from the corresponding property of $h_{X,\ell}$. As an application, we have :

**Proposition 7.7.** *There exist smooth projective varieties $Y_\alpha$ and $Z_\beta$, of dimension $\leqslant \dim X_0$, such that $h^i_{X,\ell} = \sum_\alpha h^i_{Y_\alpha,\ell} - \sum_\beta h^i_{Z_\beta,\ell}$ for every $i$ and every prime $\ell$.*

*Proof.* This is done by induction on $\dim X_0$. The method is the same as the one used for Proposition 7.3 ; if $U \subset X$ and $Z \supset U$ are as in that proof, the induction assumption gives us a decomposition of the $h^i_{Z-U,\ell}$ and $h^i_{X-U,\ell}$ of the required type. By additivity, we have :

$$h^i_{X,\ell} = h^i_{X-U,\ell} - h^i_{Z-U,\ell} + h^i_{Z,\ell},$$

as wanted.

[2] *Warning.* The letter $i$ in $h^i_{X,\ell}$ is merely an upper index ; it should not be confused with a power. When we need the square of $h_{X,\ell}$, we shall write it $h_{X,\ell}.h_{X,\ell}$.

(Another way to present this proof is to introduce the Grothendieck group of all reduced **Q**-varieties and show that it is generated by the classes of the smooth projective varieties, cf. [GS 96].)

**Theorem 7.8**. *Assume* $X_0 \neq \varnothing$; *let* $d = \dim X_0$.

a) $h^i_{X,\ell} = 0$ *if* $i > 2d$.

b) *If* $d < i \leqslant 2d$, *then* $h^i_{X,\ell}$ *is divisible by* $\chi_\ell^{d-i}$, *i.e. its product by* $\chi_\ell^{i-d}$ *is a virtual character of weight* $2d - i$.

c) *If the finite set* $S$ *of* §7.1.1 *is large enough, then* $h^i_{X,\ell}(g_p^e)$ *belongs to* $\mathbf{Z}[1/p]$ *for every* $p \notin S$ *and every* $e$; *it belongs to* $\mathbf{Z}$ *if* $e \geqslant 0$.

d) *If* $c$ *denotes complex conjugation, the value of* $h^i_{X,\ell}$ *on* $c$ *is an integer that does not depend on* $\ell$; *it is* 0 *if* $i$ *is odd.*

e) *Let* $\ell'$ *be another prime number. If the set* $S$ *is large enough, and if* $p$ *is a prime with* $p \notin S$, *the values of* $h^i_{X,\ell}$ *and* $h^i_{X,\ell'}$ *on the elements* $g_p^e$ $(e \in \mathbf{Z})$ *are the same.*

[Note that property e) makes sense because of c).]

*Proof.* By Proposition 7.7, it is enough to prove this when $X_0$ is projective, smooth, and all its components have dimension $d' \leqslant d$. The case $d' < d$ is a consequence of the case $d' = d$. We may thus assume that $d' = d$. We then have $h^i_{X,\ell} = h_{i,X,\ell}$ and the theorem follows from the well-known properties of the $\ell$-adic cohomology of projective smooth varieties (including Deligne's theorem). More precisely :

a) : clear.

b) : Suppose $d < i \leqslant 2d$; let $L \in H^2(X_0, \ell) \otimes \mathbf{Q}_\ell(1)$ be the $\ell$-adic cohomology class of an ample divisor of $X_0$. Hodge theory shows that the cup-product by the $(i-d)$-power of $L$ gives an isomorphism

$$H^{2d-i}(X_0, \ell) \otimes \mathbf{Q}_\ell(d-i) \ \tilde{\rightarrow}\ H^i(X_0, \ell).$$

This isomorphism is compatible with the action of $\Gamma_{\mathbf{Q}}$. Hence $h^i_{X,\ell}$ is equal to $\chi_\ell^{d-i} h'$, with $h' = h^{2d-i}_{X,\ell}$.

c) and e) : follow from Deligne's Theorem 4.5.

d) : follows from Artin's comparison theorem and the fact that complex conjugation interchanges the summands of the Hodge decomposition of the cohomology.

A corollary of e) is that, for a given $i$, the value at 1 of $h^i_{X,\ell}$ belongs to $\mathbf{Z}$ and is independent of $\ell$. We shall denote it by $B^i(X)$, or $B^i(X_0)$. It is the *i-th virtual Betti number* of $X_0$. If $Y_\alpha$ and $Z_\beta$ are chosen as in

Proposition 7.7, one has

$$B^i(X) = \sum_\alpha B_i(Y_\alpha) - \sum_\beta B_i(Z_\beta).$$

Note the formula :

$$\sum(-1)^i B^i(X_0) = h_X(1) = \chi(X(\mathbf{C})) = \sum(-1)^i B_i(X),$$

where now $\chi$ denotes the Euler-Poincaré characteristic (instead of the cyclotomic character ...).

*Remarks.*

1) The weight decomposition of the cohomology can be refined into a decomposition in the setting of *Chow motives*, cf. [GS 96]. Indeed, one of the reasons for Grothendieck introducing motives in 1964 was to explain the existence of the virtual Betti numbers, which I had pointed out to him as an intriguing consequence of Weil's conjectures.

2) Part e) of Theorem 7.8 expresses the *compatibility* (in the sense of [Se 68, I.2.3]) of the characters $h^i_{X,\ell}$ for fixed $X, i$, when $\ell$ runs through the different primes. As a matter of fact, the proof also gives *strong compatibility*, namely the existence of a finite set $S$ of primes such that " everything behaves well outside $S$ " in the following sense :

(7.1.2.1) Each $h^i_{X,\ell}$ is of weight $i$ outside $S_\ell = S \cup \{\ell\}$.

(7.1.2.2) If $\ell, \ell'$ are two primes, and if $p \notin S_{\ell,\ell'} = S \cup \{\ell\} \cup \{\ell'\}$, the values of $h^i_{X,\ell}$ and $h^i_{X,\ell'}$ at all the $g_p^e$ are the same.

[*Proof.* Choose $S$ such that the $\mathbf{Q}$-varieties $Y_\alpha$ and $Z_\beta$ of Proposition 7.7 have good reduction outside $S$, i.e. are the generic fibers of smooth projective schemes over $\operatorname{Spec} \mathbf{Z} - S$. ]

Property (7.1.2.2) allows us to introduce the notation $h^i_X(p^e)$ for $p \notin S$, and $e \in \mathbf{Z}$, as the common value of the $h^i_{X,\ell}(g_p^e)$, for all $\ell \neq p$. By the results of Katz-Laumon[3] quoted in §4.8.4, one may choose $S$ such that

(7.1.2.3) $N_X(p^e) = \sum_i (-1)^i h^i_X(p^e)$ for every $p \notin S$, and every $e$.

The decomposition of $N_X(p)$ so obtained is convenient for describing its archimedean properties since each summand $h^i_X(p^e)$ is $O(p^{ei/2})$ if $e > 0$ and is expected (see Chapter 8) to be often of that size, unless it is identically 0. In the next section we are going to look more closely into the two top terms $i = 2d$ and $i = 2d-1$, together with the lowest term $i = 0$.

[3]This can also be done, more simply, by reduction to the case where $X_0$ is projective and smooth.

*The zeta point of view.*

Let $p$ be a prime, and let $\zeta_{X,p}(s)$ be the $p$-component of $\zeta_X(s)$, cf. §1.5. By Deligne's Theorem 4.5, we may write it as follows :

$$\zeta_{X,p}(s) = \prod_j (1-\alpha_j p^{-s}) / \prod_l (1-\beta_l p^{-s}),$$

where the $\alpha_j$ and the $\beta_l$ are $p$-Weil integers. We have

$$(*) \quad N_X(p^e) = \sum_l \beta_l^e - \sum_j \alpha_j^e \text{ for every } e \in \mathbf{Z}.$$

We may rewrite this formula by putting together all the terms with the same weight. This gives :

$$N_X(p^e) = \sum_i N_X^i(p^e),$$

where $N_X^i(p^e)$ is defined by the same formula as (*), the sums being restricted to the $\alpha_j$ and the $\beta_l$ that have weight $i$. *For $p$ large enough, $N_X^i(p^e)$ coincides with the rational number $h_X^i(p^e)$ defined above if $i$ is even, and with $-h_X^i(p^e)$ if $i$ is odd.* This is easy to check when $X_0$ is a projective smooth variety, and the general case follows by reduction to that case.

*Exercise.* Show that $h_X^i(p^{-e}) = p^{-ie} h_X^i(p^e)$ for every $e \in \mathbf{Z}$; if moreover $X_0$ is smooth projective, and all its components have dimension $d$, show that $h_X(p^{-e}) = p^{-de} h_X(p^e)$.

[Hint. Use the exercise of §7.1.2.]

## 7.2. The weight decomposition : examples and applications

We keep the hypotheses and notation of Theorem 7.8; in particular, we assume $X_0 \neq \varnothing$, and we put $d = \dim X_0$.

### 7.2.1. The dominant term : $i = 2d$

**Lemma 7.9.** *If $i = 2d$ or $2d-1$, the virtual character $h_{X,\ell}^i$ is effective, and it is a birational invariant of $X_0$.*

*Proof.* The birational invariance means that $h_{X,\ell}^i$ does not change when one adds to $X_0$, or removes from it, a subvariety $Z$ of dimension $\leqslant d-1$; this follows from the additivity of $h_{X,\ell}^i$ and the vanishing of $h_{Z,\ell}^i$ for $i > 2\dim Z$. By Hironaka's theorem, there exists a projective smooth variety $Y$ of dimension $\leqslant d$ that is birationally equivalent to $X_0$ (assuming $X_0$ to be reduced, which is no restriction). Hence we have

$$h_{X,\ell}^i = h_{Y,\ell}^i = h_{i,Y,\ell} \text{ when } i = 2d \text{ or } 2d-1,$$

which shows that these characters are effective.

The case $i = 2d$ is easy to describe concretely :

Let $I_X$ be the set of irreducible components of dimension $d$ of $\overline{X} = X_{/\overline{\mathbf{Q}}}$. There is a natural action of $\Gamma_{\mathbf{Q}}$ on $I_X$. Let $\varepsilon_X : \Gamma_{\mathbf{Q}} \to \mathbf{N}$ be the permutation character of that action ; if $\gamma \in \Gamma_{\mathbf{Q}}$, we have

$$\varepsilon_X(\gamma) = |I_X^\gamma| = \text{ number of elements of } I_X \text{ fixed by } \gamma.$$

**Proposition 7.10.** $h^{2d} = \chi_\ell^{-d}\varepsilon_X$, *i.e.* $h^{2d}(p) = p^d\varepsilon_X(g_p)$ *for every large enough* $p$. *In particular, we have* $B^{2d}(X) = |I_X|$.
[Note that $\varepsilon_X(g_p) = \varepsilon_X(\sigma_p)$ since a permutation representation is self-dual ; we shall denote this integer by $\varepsilon_X(p)$.]

*Proof.* Thanks to Proposition 7.7, we may assume that $X_0$ is projective, smooth, and that all its components have dimension $d$. Poincaré duality then shows that

$$h^i_{X,\ell} = \chi_\ell^{d-i} h^{2d-i}_{X,\ell}.$$

For $i = 2d$, this gives what we want [4].

**Corollary 7.11.** *We have* $N_X(p) = \varepsilon_X(p)p^d + O(p^{d-\frac{1}{2}})$ *when* $p \to \infty$.

Indeed, the other terms of formula (7.1.2.3) are $O(p^{d-\frac{1}{2}})$ since they come from virtual characters of weight $\leqslant 2d-1$. Note that there is an analogous result for $N_X(p^e)$, namely :

$$|N_X(p^e) - \varepsilon_X(p^e)p^{ed}| \ll p^{e(d-\frac{1}{2})}, \quad \text{for } p \text{ large enough, and } e \geqslant 1,$$

where $\varepsilon_X(p^e)$ stands for $\varepsilon_X(g_p^e)$.

**Corollary 7.12.** *Let* $r = |I_X|$ *be the number of irreducible components of dimension* $d$ *of* $\overline{X}$. *We have*

$$\limsup N_X(p)/p^d = r.$$

*Moreover, there exists a frobenian set of primes* $P_X$, *of density* $\geqslant 1/r!$, *such that*

$$|N_X(p) - rp^d| \ll p^{d-\frac{1}{2}} \quad \text{for } p \in P_X.$$

*Proof.* The lim sup assertion follows from the previous corollary, and the fact that $0 \leqslant \varepsilon_X(p) \leqslant \varepsilon_X(1) = r$. As for $P_X$, one takes the set of primes $p$ that act trivially on $I_X$ ; this is a frobenian set, whose density is $1/|G|$, where $G$ is the image of $\Gamma_{\mathbf{Q}} \to \mathrm{Aut}(I_X)$ ; the order $|G|$ of $G$ is a divisor of $r!$ .

[4] Another proof, independent of the resolution of singularities, follows from the general properties of the "trace map", cf. [SGA 4, XVIII.2.9] or [Mi 80, VI §6].

**Corollary 7.13**. *Let $r_0$ be the number of* $\mathbf{Q}$*-irreducible components of $X_0$ of dimension d. There exists $c > 0$ such that :*

$$\sum_{p \leqslant x} N_p(X) = r_0 \mathrm{Li}(x^{d+1}) + O(x^{d+1} e^{-c\sqrt{\log x}}) \quad \textit{for } x \to \infty.$$

*Under GRH, the O-term on the right can be replaced by $O(x^{d+\frac{1}{2}} \log x)$.*

[In particular, we have $\sum_{p\leqslant x} N_p(X) \sim \frac{r_0}{d+1} x^{d+1}/\log x$ for $x \to \infty$.]

*Proof.* By Corollary 7.11, we may write the left hand term of the formula as

$$\sum_{p\leqslant x} \varepsilon(p) p^d + O\left(\sum_{p\leqslant x} p^{d-\frac{1}{2}}\right).$$

By Theorem 3.5, applied to $m = d - \frac{1}{2}$, the right hand term is $O(x^{d+\frac{1}{2}}/\log x)$; it can thus be neglected. As for the left hand term, note first that $r_0$ is the number of $\Gamma_{\mathbf{Q}}$-orbits of $I_X$, i.e. the *mean value* of $\varepsilon_X$ in the sense of §3.3.3.5 ; as explained there, Theorem 3.5, applied to $f(p) = \varepsilon_X(p)$ and to $m = d$, gives :

$$\sum_{p\leqslant x} \varepsilon_X(p) p^d = r_0 \mathrm{Li}(x^{d+1}) + O(x^{d+1} e^{-c\sqrt{\log x}}) \quad \text{for } x \to \infty.$$

The corollary follows.

**Corollary 7.14**. *Suppose that $X_0$ is normal and that all its irreducible components have dimension d. Then, if p is large enough, we have :*

$$N_p(X) > 0 \iff \varepsilon_X(p) > 0.$$

[Recall that $\varepsilon_X$ is the permutation character associated with the set $I_X$, and that $\varepsilon_X(p)$ is an abbreviation for $\varepsilon_X(g_p) = \varepsilon_X(\sigma_p)$.]

*Proof.* If $\varepsilon_X(p) > 0$, Corollary 7.11 shows that $N_X(p) \geqslant p^d + O(p^{d-\frac{1}{2}})$, hence $N_X(p) > 0$ if $p$ is large enough.

It remains to show that $N_X(p) = 0$ if $\varepsilon_X(p) = 0$ and $p$ is large enough. We can obviously remove from $X$ a finite number of $X_p$ ; this allows us to assume that $X$ is normal, and also that it is affine and irreducible. Let $A = H^0(X, \mathcal{O}_X)$ be the affine ring of $X$ ; it is a normal domain. Let $K$ be the field of fractions of $A$ (i.e. the field of rational functions on $X_0$) and let $E$ be the algebraic closure of $\mathbf{Q}$ in $K$, i.e. the maximal subfield of $K$ that is algebraic over $\mathbf{Q}$. The field $E$ is a finite extension of $\mathbf{Q}$ (see e.g. [A IV-VII, Chap.V, §14, cor.1 to prop.17]). Let $O_E$ be the ring of integers of $E$ ; since the elements of $O_E$ are integral over $\mathbf{Z}$, and $A$ is normal, $O_E$ is contained in $A$; hence $E$ is contained in $A_0 = \mathbf{Q} \otimes A$, which is the affine ring of $X_0$. Let $I_E$ be the set of all embeddings $E \to \overline{\mathbf{Q}}$; this is a $\Gamma_{\mathbf{Q}}$-set, that is isomorphic to the set denoted previously by $I_X$. [Indeed, every embedding $\iota : E \to \overline{\mathbf{Q}}$ gives a $\overline{\mathbf{Q}}$-algebra $A_\iota = A \otimes_E \overline{\mathbf{Q}}$, and the $\overline{\mathbf{Q}}$-algebra $\overline{A} = A \otimes_{\mathbf{Q}} \overline{\mathbf{Q}}$ is the direct product of the $A_\iota$; hence the minimal ideals of $\overline{A}$ correspond bijectively to the embeddings $E \to \overline{\mathbf{Q}}$.] Let $S_E$ be the set of primes which divide the discriminant of $E$. If $p \notin S_E$, the ring $O_E/pO_E$ is a product of finite extensions of $\mathbf{F}_p$; let $n_1(p), ..., n_k(p)$ denote the degrees of these extensions. These degrees are equal to the orders of the orbits of $\sigma_p$ on $I_E$. If $\varepsilon_X(p) = 0$, none of these orbits has order 1. This means that there does not exist any homomorphism $O_E \to \mathbf{F}_p$; since $O_E$ is a subring of $A$, there does not exist any homomorphism $A \to \mathbf{F}_p$, i.e. $N_X(p)$ is 0.

*Exercise.* Suppose that $X_0$ is normal and $\mathbf{Q}$-irreducible, but not geometrically irreducible, so that the number $r$ of its $\overline{\mathbf{Q}}$-irreducible components is $> 1$. Show that the density of the set of $p$ with $N_X(p) = 0$ is $\geqslant 1/r$.

[Hint. Use Corollary 7.14 together with the following result of P.J. Cameron and A.M. Cohen ([CC 92], see also [Se 02]) : if $G$ is a finite group acting transitively on a set with $r$ elements, with $r > 1$, and if $R$ is the set of $g \in G$ which have no fixed point, then $|R|/|G| \geqslant 1/r$.]

### 7.2.2. The next-to-dominant term : $i = 2d - 1$

As shown in Lemma 7.9, the character $h^{2d-1}_{X,\ell}$ is effective, and is a birational invariant of $X_0$. To describe it, we may thus assume that $X_0$ is a smooth projective variety over $\mathbf{Q}$, and we may also assume that it is $\mathbf{Q}$-irreducible. As in the section above, let $K$ be its function field, and let $E$ be the maximal subfield of $K$ that is algebraic over $\mathbf{Q}$. The field $E$ is a finite extension of $\mathbf{Q}$, and the projection $X_0 \to \operatorname{Spec} \mathbf{Q}$ factors in

$$X_0 \;\to\; \operatorname{Spec} E \;\to\; \operatorname{Spec} \mathbf{Q}.$$

We may thus view $X_0$ as a geometrically irreducible smooth projective variety over $E$ (this is an example of a "Stein factorization", cf. [Ha 77, Cor.11.5]).

Let *Alb* denote the Albanese variety of the $E$-variety $X_0$; it is an abelian variety over $E$. Let $A = R_{E/\mathbf{Q}} Alb$ be the $\mathbf{Q}$-abelian variety deduced from *Alb* by the Weil's restriction of scalars functor (see e.g. [We 61, §1.3] and [BLR 90, §7.6]). This variety gives us the character $h^{2d-1}_{X,\ell}$ we were looking for. More precisely :

**Proposition 7.15.** *We have* $h^{2d-1}_{X,\ell} = \chi_\ell^{1-d}\, h^1_{A,\ell}$. *In particular* $B^{2d-1}(X) = 2\dim A$ .

[Recall that $h^1_{A,\ell}$ is the character of $\Gamma_{\mathbf{Q}}$ associated with the first cohomology group of $A_{/\overline{\mathbf{Q}}}$, i.e. the dual of the $\ell$-Tate module of $A$.]

*Proof.* This follows from the following two facts from Hodge theory, combined with the standard construction of the Albanese variety :

a) There is a natural isomorphism $H^1(A,\ell) \to H^1(X_0,\ell)$.

b) One has $h^{2d-1}_{X,\ell} = \chi_\ell^{1-d}\, h^1_{X,\ell}$, cf. proof of Proposition 7.10.

*Remark.* Instead of using the Albanese variety, we could have used the Picard variety : these varieties are isogenous, hence have the same $\ell$-adic cohomology.

*Example.* Suppose $E = \mathbf{Q}$, i.e. $X_0$ is geometrically irreducible. In that case, Proposition 7.14 gives the following description of the integer $h^{2d-1}(p)$, when $p$ is large enough : it is equal to $p^{d-1}\mathrm{Tr}(F_p)$, where $F_p$ is the Frobenius endomorphism of $A_p$. If $g = \frac{1}{2}B^{2d-1}(X)$ is the dimension of $A$, we have

$$-2gp^{d-\frac{1}{2}} \leqslant h^{2d-1}(p) \leqslant 2gp^{d-\frac{1}{2}}.$$

It is expected that these bounds are essentially optimal, i.e. that the ratio $h^{2d-1}(p)/p^{d-\frac{1}{2}}$ can be arbitrary close to any given point of the interval $[-2g, 2g]$. This would follow from the general Sato-Tate conjecture; see Chapter 8 for more precise statements (esp. §8.1.4.2) and for references to the case $g = 1$, which has been settled recently (see §8.1.5).

### 7.2.3. The lowest term : $i = 0$

**Proposition 7.16.** *The virtual character $h^0_{X,\ell}$ is a* $\mathbf{Z}$*-linear combination of permutation characters; it is independent of the choice of $\ell$.*

*Proof.* By Proposition 7.7, it is enough to prove this when $X_0$ is a smooth $\mathbf{Q}$-irreducible projective variety; in that case, $h^0_{X,\ell}$ is equal to the permutation character $\varepsilon_X$ defined in §7.2.1, namely the character associated with the action of $\Gamma_{\mathbf{Q}}$ on the set of irreducible components of the $\overline{\mathbf{Q}}$-variety $\overline{X} = X_{0/\overline{\mathbf{Q}}}$.

*Remark.* Proposition 7.16 implies that $h^0_{X,\ell}$ is a virtual character *over* $\mathbf{Q}$, i.e. the difference of two continuous linear representations $\Gamma_{\mathbf{Q}} \to \mathbf{GL}_n(\mathbf{Q})$ and $\Gamma_{\mathbf{Q}} \to \mathbf{GL}_m(\mathbf{Q})$, for suitable integers $n, m$, with $n - m = B^0(X)$. We may thus write it simply $h^0_X$, without mentioning $\ell$. The fact that this character is a difference of two permutation characters is a non-trivial piece of information : there exist characters over $\mathbf{Q}$ which are not of that form, cf. [Se 77, §13.1, exerc. 4].

*Exercises.*

1) Let $\psi : \Gamma_{\mathbf{Q}} \to \mathbf{Z}$ be a virtual character of $\Gamma_{\mathbf{Q}}$ that is the difference of two continuous permutation characters. Show that there exists $X$ with $h^0_X = \psi$.
[Hint. It is enough to prove this when $\psi$ is a permutation character (take $X_0$ of dimension 0 in that case), and when $-\psi$ is a permutation character (take $X_0 = \mathbf{A}^1 - Y$, where $Y$ is a 0-dimensional closed subscheme of $\mathbf{A}^1$).]

2) Let $G = \mathbf{PSL}_2(\mathbf{F}_5)$ or $\mathbf{PSL}_2(\mathbf{F}_7)$. Show that every $\mathbf{Q}$-valued virtual character of $G$ is a difference of two permutation characters. (This applies in particular to the virtual characters with set of values $\{0, 3\}$ constructed in Exercise 2 of §5.1.5 ; by Exercise 1 above, these characters can occur as $h^0_X$ for a suitable $X$.)

3) Let $X, Y$ be two schemes of finite type over $\mathbf{Z}$. Use Theorem 7.1 to show that the following four properties are equivalent :

a) $|N_X(p) - N_Y(p)|$ remains bounded when $p$ varies, cf. §6.3.1.

b) $h_{X,\ell} - h_{Y,\ell}$ has weight 0.

c) $h_{X,\ell} - h_{Y,\ell}$ is the difference of two permutation characters.

d) $h^i_{X,\ell} = h^i_{Y,\ell}$ for every $i > 0$ (for every $\ell$ – or for a given $\ell$, it amounts to the same).

If these properties hold, show that there exist two polynomials $f, g \in \mathbf{Z}[t]$ such that $N_X(p^e) + N_f(p^e) = N_Y(p^e) + N_g(p^e)$ for every large enough $p$ and every $e$.

### 7.2.4. The set of $p$ with $N_X(p) > 0$

Let us denote by $P_X$ the set of $p \in P$ such that $N_X(p) > 0$. In the special case $X = \operatorname{Spec} \mathbf{Z}[t]/(f)$, with $f \in \mathbf{Z}[t]$, let us write $P_f$ instead of $P_X$, so that $p \in P_f \iff f$ has a root mod $p$.

The following theorem shows that every $P_X$ is equal to some $P_f$ : the special case gives as much as the general case. This surprising result is essentially due to J. Ax and L. van den Dries ([Ax 67], [Dr 91]). More precisely :

**Theorem 7.17.** *Let $Q$ be a subset of $P$. The following properties are equivalent :*

i) *There exists a scheme $X$ of finite type over $\mathbf{Z}$ such that $Q = P_X$.*

ii) *There exists $f \in \mathbf{Z}[t]$ such that $Q = P_f$.*

iii) *There exists a continuous permutation character $\varepsilon$ of $\Gamma_{\mathbf{Q}}$ such that $\varepsilon(g_p) > 0 \iff p \in Q$ for every large enough $p$.*

iv) *The set $Q$ is frobenian, and if $U_Q$ is the corresponding open and closed subset of* $\Gamma_{\mathbf{Q}}$ (in the sense of §3.3.1), *then $U_Q$ is stable under $\gamma \mapsto \gamma^e$ for every $e \in \mathbf{Z}$.*

*Proof.* Let us show first that i), ii) and iii) are equivalent :

ii) $\Rightarrow$ i) is clear.

iii) $\Rightarrow$ ii). Let $I$ be a finite set with a continuous $\Gamma_{\mathbf{Q}}$-action with character $\varepsilon$. By Galois theory there exists a finite étale $\mathbf{Q}$-algebra $E$ such that $\mathrm{Hom}_{alg}(E, \overline{\mathbf{Q}}) \simeq I$, see e.g. [A IV-VII, §V.10.10]. Since $\mathbf{Q}$ is infinite, $E$ can be generated by one element ([A IV-VII, §V.7, prop.7]). This means that we can find a non-zero polynomial $h \in \mathbf{Q}[t]$ such that $E \simeq \mathbf{Q}[t]/(h)$. Since we are free to multiply $h$ by a non-zero scalar, we may assume that $h$ has coefficients in $\mathbf{Z}$. Let $n = [E : \mathbf{Q}] = \deg h$. Since $E$ is étale, the discriminant $d$ of $h$ is $\neq 0$. Let $S$ be the set of primes that divide either $d$ or the highest coefficient of $h$. If $p \notin S$, the $\mathbf{F}_p$-algebra $\mathbf{F}_p[t]/(h)$ is étale of rank $n$, and it splits into a product of extensions of $\mathbf{F}_p$ whose degrees are the orders of the orbits of $\sigma_p$ (or of $g_p$) acting on $I$. Hence, for $p \notin S$, we have $\varepsilon(g_p) > 0$ if and only if $h$ has a zero mod $p$. By iii), this means that the two sets $P_h$ and $Q$ only differ by a finite set. It remains to replace $h$ by another polynomial $f$ in such a way that $P_f$ and $Q$ actually coincide; this follows from the following lemma, which shows that one can add or substract any given prime to $P_h$ :

**Lemma 7.18.** *Let $h$ be a non-zero element of $\mathbf{Z}[t]$, and let $p$ be a prime number. There exist non-zero elements $h_+$ and $h_-$ of $\mathbf{Z}[t]$ with the following properties :*

a) *$h_+$ has a root* mod $p$;

b) *$h_-$ has no root* mod $p$;

c) *If $p' \neq p$, then $h_+$ and $h_-$ have a root* mod $p'$ *if and only if $h$ does.*

[More briefly : $P_{h_+} = P_h \cup \{p\}$, $P_{h_-} = P_h$ if $p \notin P_h$ and $P_{h_-} = P_h - \{p\}$ if $p \in P_h$.]

*Proof of Lemma 7.18.* Put $h_+ = ph$ and define $h_-$ as follows :

if the constant term of $h$ is 0, take $h_- = 1 + pt$;

if the constant term of $h$ is $\neq 0$, let $e$ be its $p$-adic valuation and take $h_-(t) = p^{-e}h(p^{e+1}t)$.

*Proof of* i) $\Rightarrow$ iii). Let $X$ be of finite type over $\mathbf{Z}$. We want to prove that there exists a permutation character $\varepsilon$ such that $\varepsilon(g_p) > 0 \Longleftrightarrow N_X(p) > 0$ for every large enough $p$. If $X$ is an union of two closed (or open) subschemes and if the statement is true for both of these, then it is true for $X$. By a standard argument, we may thus assume that $X$ is normal and irreducible, in which case the result follows from Corollary 7.14.

It remains now to see that property iv) is equivalent to i), ii), iii) :

iii) $\Rightarrow$ iv). If iii) holds, it is clear that $Q$ is frobenian. The stability of $U_Q \subset \Gamma_{\mathbf{Q}}$ by $\gamma \mapsto \gamma^e$ is also clear : if $g_p$ has a fixed point, the same is true for all the $g_p^e$.

iv) $\Rightarrow$ iii). Assume iv). Choose a normal open subgroup $N$ of $\Gamma_{\mathbf{Q}}$ such that $N.U_Q = U_Q$, and let $G = \Gamma_{\mathbf{Q}}/N$; let $U_N$ be the image of $U_Q$ in $G$. The set $U_N$ is stable under conjugation, and under the power maps $\gamma \mapsto \gamma^e$, $e \in \mathbf{Z}$. We are thus reduced to proving the following lemma :

**Lemma 7.19.** *Let $G$ be a finite group, and let $V$ be a subset of $G$ that is stable under conjugation and under the power maps $g \mapsto g^e$. There exists a finite $G$-set $I$ such that $g \in V \iff I^g \neq \varnothing$.*

*Proof of Lemma 7.19.* Let $\mathcal{C}$ be the set of cyclic subgroups of $G$ contained in $V$. Take for $I$ the disjoint union of the $G/C$ for $C \in \mathcal{C}$.

This concludes the proof of Theorem 7.17.

*Exercises.*

1) Show that, in property iv) of Theorem 7.17, one may replace " for every $e \in \mathbf{Z}$ " by " for every $e$ in a subset of $\mathbf{Z}$ that is dense in $\hat{\mathbf{Z}}$ ".

2) Let $X$ be a scheme of finite type over $\mathbf{Z}$, let $e$ be an integer $\geqslant 1$ and let $P_X^e$ be the set of primes $p$ such that $N_X(p^e) > 0$. Show that $P_X^e$ has the four properties listed in Theorem 7.17.

[Hint. If $X$ is quasi-projective, its cyclic $e$-th power $X_e = X^e/C_e$ is well defined ($C_e$ being the cyclic group $\mathbf{Z}/n\mathbf{Z}$ acting on $X^e = X \times \cdots \times X$ by cyclic permutations of the $e$ factors). Show that $P_X^e = P_{X_e}$.]

### 7.2.5. Application : a characterization of affine-looking schemes

As above we assume $X_0 \neq \varnothing$; let $d$ be its dimension.

Let us say that $X_0$ is "affine-looking" if $N_X(p) = p^d$ for all large enough $p$ . This is equivalent to $h_{X,\ell} = \chi_\ell^{-d}$, and hence also to $N_X(p^e) = p^{ed}$ for all large enough $p$ and all $e \in \mathbf{Z}$. Note that, in such a case, we have the identity

$$N_X(p^e) = N_X(p)^e \quad \text{for all large enough } p \text{ and all } e.$$

Conversely, let us show that, if this equation holds for $e = 2$, then $X_0$ is affine-looking. More precisely :

**Theorem 7.20.** *Suppose that $N_X(p^2) = N_X(p)^2$ for every $p$ in a set of density 1. Then $X_0$ is affine-looking.*

*Proof.* Choose a prime $\ell$, and a finite set $S$ of primes such that $N_X(p^e) = h_{X,\ell}(g_p^e)$ for every $p \notin S$. Let us drop the indexes $X, \ell$ from the notation. We have :

$$N_X(p^2) = h(g_p^2) = \Psi^2 h(g_p) \ \text{ if } p \notin S.$$

[Recall that $\Psi^2 h$ is the function $\gamma \mapsto h(\gamma^2)$, cf. §5.1.1.3.]

By assumption, the two virtual characters $\Psi^2 h$ and $h.h$ have the same value for all the $g_p$ where $p$ runs through a set of density 1. Since such $g_p$ are dense in $\mathrm{Cl}\,\Gamma_S$, this implies that $\Psi^2 h = h.h$. Let $h^{2d}$ be the component of weight $2d$ of $h$; the components of weight $4d$ of $\Psi^2 h$ and $h.h$ are respectively $\Psi^2 h^{2d}$ and $h^{2d}.h^{2d}$. Hence we have

$$(*) \quad \Psi^2 h^{2d} = h^{2d}.h^{2d}.$$

Let us write $h^{2d} = \chi^{-d}\varepsilon$ as in Proposition 7.9, where $\chi$ is the cyclotomic character and $\varepsilon = \varepsilon_X$ is a permutation character. We have $\Psi^2 h^{2d} = \chi^{-2d}\Psi^2\varepsilon$, hence formula $(*)$ shows that $\Psi^2\varepsilon = \varepsilon.\varepsilon$. Let $n = \varepsilon(1)$ be the degree of the permutation character $\varepsilon$; the value at 1 of $\Psi^2\varepsilon$ is $n$ and the value at 1 of $\varepsilon.\varepsilon$ is $n^2$. Hence $n^2 = n$, which implies $n = 1$ since $n > 0$; the main term of $h$ is thus $\chi^{-d}$.

Let us now show that all the other terms of the weight decomposition of $h$ are 0. Suppose not, and let $h^i$ be the highest non-zero one, with $0 \leqslant i < 2d$. We have

$$h = \chi^{-d} + h^i + \text{ terms of weight } < i,$$

which implies :

$$h.h = \chi^{-2d} + 2\chi^{-d}h^i + \text{ terms of weight } < 2d + i,$$

and

$$\Psi^2 h = \chi^{-2d} + \Psi^2 h^i + \text{ terms of weight } < 2i.$$

In particular, the component of weight $2d+i$ of $\Psi^2 h$ is 0. By comparing with the expansion of $h.h$, we get $\chi^{-d}h^i = 0$ and since $\chi^{-d}$ is invertible, this implies $h^i = 0$, which is a contradiction. Hence $h = \chi^{-d}$, as wanted.

*Remarks.*

1) The relation $\Psi^2 h = h.h$ is equivalent to $\lambda^2 h = 0$, and the proof above could have been written in terms of $\lambda^2$ instead of $\Psi^2$; one would then use the identity $\lambda^2(x+y) = \lambda^2 x + \lambda^2 y + xy$.

2) The assumption that the set of $p$ has density 1 can be weakened, cf. exercise 2 below.

3) Suppose $X_0$ is affine-looking; it then has only one irreducible component of dimension $d$, that we assume to be reduced. One may ask *whether that component is birationally isomorphic to the affine space* $\mathbf{A}^d$, i.e. whether its function field is isomorphic to $\mathbf{Q}(t_1, ..., t_d)$. The answer is : *no.*

Indeed, let $G$ be a finite subgroup of $\mathbf{GL}_d(\mathbf{Q})$; choose for $X_0$ the quotient $\mathbf{A}^d/G$. As mentioned in §6.1.3, $X_0$ is affine-looking; its function field is the subfield of $\mathbf{Q}(t_1, ..., t_d)$ fixed under $G$ and there are well-known examples where this field is not purely transcendental over $\mathbf{Q}$, for instance when $G$ is cyclic of order 47 ([Sw 69], [Vo 70]) or of order 8 ([Le 74]).

*Exercises.*

1) Let $a, b$ be integers $> 1$ and suppose that $N_X(p^a) = N_X(p)^b$ for every $p$ in a set of density 1. Show that $X_0$ is affine-looking, and that its dimension is 0 if $a \neq b$.

[Hint. Same method as for Theorem 7.20.]

2) Let $B = \sum_i B_i(X)$, and $C = B(B-1)/2$. Show that Theorem 7.20 remains valid if " density 1 " is replaced by " density $> 1 - 1/C^2$ ".

[Hint. Use exerc. 3 of §5.3.3.]

# CHAPTER 8

# THE SATO-TATE CONJECTURE

The original Sato-Tate conjecture (which recently became a theorem) was about an elliptic curve $X$ over $\mathbf{Q}$, with no complex multiplication : if the corresponding $N_X(p)$ is written as $p+1-a_p$, it said that the ratio $a_p/p^{\frac{1}{2}}$ is equidistributed in the interval $[-2,2]$ with respect to a measure which is independent of $X$, namely the Sato-Tate measure, cf. §8.1.5.2. This conjecture has a natural generalization to every motive ([Se 94, §13]).

The aim of the present chapter is to explain this general conjecture in the context of $N_X(p)$, for an arbitrary $X$, and to see what it implies concretely. As in the previous chapter, we assume that the ground field is $\mathbf{Q}$, but almost everything could be done over any field which is finitely generated over $\mathbf{Q}$, as explained in §9.5.4.

## 8.1. Equidistribution statements

We start by giving a list of statements which are consequences of the general Sato-Tate conjecture.

### 8.1.1. Introduction

Let $X$ be, as usual, a scheme of finite type over $\mathbf{Z}$. For every $w \in \mathbf{N}$, we have defined in §7.1 an $\ell$-adic virtual character $h^w_{X,\ell}$ that is of weight $w$ outside some finite set $S$ of primes; if $p \notin S$, let us denote by $h^w(p)$ (or $h^w_X(p)$ if we want to keep track of $X$) the value of that character at the geometric Frobenius $g_p$;

$$N_X(p) = \sum_w (-1)^w h^w(p)$$

and each $h^w(p)$ has absolute value $\ll p^{w/2}$; more precisely (cf. §7.1.1), if $h^w_{X,\ell}$ is the difference of two effective characters of degree $a$ and $b$, we have $|h^w(p)| \leqslant Cp^{w/2}$, with $C = a+b$. We now normalize $h^w(p)$ by defining $f(p) \in \mathbf{R}$ via the formula

$$f(p) = h^w(p)/p^{w/2}.$$

[When we need to specify $X$ and $w$, we shall write $f_{X,w}$ instead of merely $f$. We may also define in a similar way $f(p^e) = h^w(p^e)/p^{ew/2}$ for every $e \in \mathbf{Z}$, and in particular $f(1) = h_w(1) = a-b$, with the same notation as above.

Note that the case $e < 0$ is not very interesting, since it follows from §7.1.1 that $f(p^{-e}) = f(p^e)$.]

Let $I = [-C, C]$. We have $f(p) \in I$. When $p$ varies, *what is the distribution of $f(p)$ in the interval $I$?* The general form of the Sato-Tate conjecture, that we shall state more precisely in §8.2, gives a partial answer to that question. Indeed, this conjecture implies the following statements[1] (see §8.4 for the proofs) :

### 8.1.2. Equidistribution

There exists a (unique) positive measure $\mu$ of mass 1 on $I = [-C, C]$ such that the sequence $f(p)$ $(p \in P - S)$ is $\mu$-equidistributed.

Let us recall what this means. By definition (cf. [INT, Chap. II]), a positive measure $\mu$ on $I$ is a linear form $\varphi \mapsto \mu(f)$ on the space of continuous complex valued functions on $I$, with the following property :

$$\varphi \text{ real } \geqslant 0 \Rightarrow \mu(\varphi) \text{ real } \geqslant 0.$$

It is customary to write $\mu(\varphi)$ with an integral sign, such as $\int_I \varphi\mu$ or $\int_I \varphi(z)\mu(z)$ or $\int_{z\in I} \varphi(z)\mu(z)$.

The mass (or total mass) of $\mu$ is the integral of the constant function 1, i.e. $\mu(1) = \int_I \mu$.

The equidistribution of the $f(p)$'s means that

$$(*) \quad \mu(\varphi) = \lim_{x\to\infty} \frac{1}{\pi_S(x)} \sum_{p\leqslant x} \varphi(f(p)) \text{ for every continuous } \varphi,$$

where the summation is over the primes $p \notin S$ with $p \leqslant x$, and $\pi_S(x)$ is the number of such primes.

For every $z \in I$ let us denote by $\delta_z$ the Dirac measure at $z$, i.e. the linear form $\varphi \mapsto \varphi(z)$. Condition (*) means that $\mu$ is the limit, for the weak topology, of the measure $\frac{1}{\pi_S(x)} \sum_{p\leqslant x} \delta_{f(p)}$.

One can also rewrite condition (*) in terms of measures of subsets of $I$ : one asks that it be true for every characteristic function of a $\mu$-quarrable set[2], cf. [Se 68, App. to Chap.I, Proposition 1].

[1] The reader should keep in mind that the statements of §8.1.2, §8.1.3 and §8.1.4 are conditional : they depend on the Sato-Tate conjecture.

[2] Recall, cf. §6.2.1 - Corollary 6.10, that a set is called $\mu$-quarrable if its boundary has measure 0 for $\mu$. A common mistake is to apply (*) to the characteristic function of an open (or closed) subset without checking that this set is quarrable. This may be wrong, even in the simple-looking case where the subset is an interval.

### 8.1.3. Structure of $\mu$

The measure $\mu$ has remarkable properties :

8.1.3.1. *Decomposition.* It can be decomposed in a unique way as a sum $\mu = \mu^{\text{disc}} + \mu^{\text{cont}}$, where :

- The measure $\mu^{\text{disc}}$ is a finite linear combination, with $> 0$ rational coefficients, of Dirac measures $\delta_n$, associated with points $n$ of $I$; these points belong to $\mathbf{Z}$; if the weight $w$ is odd, the only possible value of $n$ is 0.
- The measure $\mu^{\text{cont}}$ has a density with respect to the Lebesgue measure. That density takes values in $[0, +\infty]$; it is continuous, integrable, and $C^\infty$ outside a finite number of points[3] of $I$.

8.1.3.2. When $w$ is odd, $\mu$ is invariant under $z \mapsto -z$.

8.1.3.3. *Support.* The support[4] of $\mu$ is the closure of the set of the $f(p)$'s. It contains all the $f(p^e)$, with $p \notin S$ and $e \in \mathbf{Z}$; in particular, it contains $f(1)$.

When $w$ is odd and $h^w_{X,\ell}$ is effective, the support of $\mu^{\text{cont}}$ is equal to the interval $[-f(1), f(1)]$.

8.1.3.4. *Integrality of the moments.* If $k$ is any integer $\geqslant 0$, let $\varphi_k$ be the function $z \mapsto z^k$ on $I$. The number $\mu(\varphi_k)$ is called the *k-th moment* of $\mu$; if we use the standard notation $\mu(\varphi) = \int_I \varphi(z)\mu(z)$, we may write it as :

$$\mu(\varphi_k) = \int_I z^k \mu(z) = \lim_{x\to\infty} \frac{1}{\pi_S(x)} \sum_{p \leqslant x} f(p)^k.$$

*All the moments of $\mu$ belong to* $\mathbf{Z}$. If $h^w_{X,\ell}$ is an effective character, *they are* $\geqslant 0$. If both $w$ and $k$ are odd, then *the k-th moment $\mu(\varphi_k)$ is* 0.

*Remark.* Note that $\mu$ *is uniquely determined by its moments*, since polynomials are dense in the Banach space of continuous functions on $I$. The same density argument shows that, if, for every $k \in \mathbf{N}$, the mean value $\frac{1}{\pi_S(x)} \sum_{p \leqslant x} f(p)^k$ has a limit when $x \to \infty$, then there exists a $\mu$ with respect to which the sequence $f(p)$ is equidistributed.

### 8.1.4. Density properties

8.1.4.1. *Inverse image of a point.* Let $z$ be a point of $I$ and let $P_z$ be the set of $p \notin S$ such that $f(p) = z$. The fact that $f(p)$ is equal to an integer

[3]These "bad" points are algebraic over $\mathbf{Q}$, but they are in general neither integral, nor rational.

[4]Recall (cf. [INT III, §2, prop.8]) that the support of $\mu$ is the smallest closed subset $Z$ such that $\varphi = 0$ on $Z \Rightarrow \mu(\varphi) = 0$.

divided by $p^{w/2}$ implies that $P_z$ has at most one element if $z \notin \mathbf{Z}$, or if $z \neq 0$ and $w$ is odd.

For the other values of $z$ (e.g. $z = 0$), $P_z$ has a density that is equal to the $\mu$-measure of the set $\{z\}$; it is $> 0$ if and only if $z$ belongs to the support of $\mu^{\text{disc}}$. Moreover, $P_z$ is the disjoint union of a frobenian set and a set of density 0 (which is sometimes infinite – see Elkies's example, end of §6.2.2).

8.1.4.2. *Inverse image of an interval.*

Let $J$ be a non-empty open interval contained in $I$, and let $P_J$ be the set of $p \notin S$ such that $f(p)$ belongs to $J$. Then $P_J$ has a density that is equal to $\mu(J)$. We have :

$$P_J \neq \varnothing \iff \text{dens}(P_J) > 0 \iff J \text{ intersects the support of } \mu.$$

As a corollary, $\text{dens}(P_J)$ is $> 0$ for every $J$ when $w$ is odd and $h^w_{X,\ell}$ is effective (provided that $C$ has been chosen equal to $f(1)$).

### 8.1.5. Example : elliptic curves

Let us consider the case where $w = 1$ and $X_0$ is an elliptic curve over $\mathbf{Q}$. For $p$ large enough, we have

$$N_X(p) = 1 - a_p + p,$$

where $a_p = h^1_X(p)$, so that $f(p) = a_p/p^{\frac{1}{2}}$ belongs to the interval $I = [-2, 2]$. Here the Sato-Tate conjecture is a theorem, see below.

There are two cases :

8.1.5.1. *CM case, cf. Example* 2.2.2. The ring $\text{End}_{\overline{\mathbf{Q}}}(X_0)$ has rank 2 over $\mathbf{Z}$. Then :

$$\mu^{\text{disc}} = \frac{1}{2}\delta_0,$$

$$\mu^{\text{cont}} = \frac{1}{2\pi}\frac{dz}{\sqrt{4 - z^2}} = \frac{1}{2\pi}d\alpha, \text{ with } z = 2\cos\alpha,\ 0 \leqslant \alpha \leqslant \pi.$$

Recall that $\delta_0$ is the Dirac measure at 0; as for $dz$, it is the restriction to $I$ of the Lebesgue measure on $\mathbf{R}$, i.e. the measure associated with the differential form $dz$ and the natural orientation of $\mathbf{R}$, cf. [FRV, §10.4.3] [5].

[5] This standard notation, which identifies differential forms with measures, is to be handled carefully because of the hidden orientations. For instance, when $z \in I$ is written as $2\cos\alpha$, with $0 \leqslant \alpha \leqslant \pi$, one has $dz = 2\sin\alpha\ d\alpha$ (as measures) instead of $dz = -2\sin\alpha\ d\alpha$ (as differential forms); similarly, $dz$, viewed as a measure, is invariant by $z \mapsto -z$, as needed for 8.1.3.2.

The $k$-th moment $\mu(\varphi_k)$ is 0 for $k$ odd; when $k$ is even and $> 0$, it is equal to $\frac{1}{2}\binom{k}{k/2}$. For $k = 0, ..., 12$ this gives [1,0,1,0,3,0,10,0,35,0,126,0,462]. One has $\mu(\varphi_k) \sim \frac{1}{\sqrt{2\pi}} 2^k/k^{\frac{1}{2}}$ when $k \to \infty$, $k$ even.

[The proof of these results uses Hecke's equidistribution theorem for Hecke characters ([He 20]), combined with the fact that the weight 1 part of the zeta function of $X$ is given by such a character, cf. [De 53].]

8.1.5.2. *non-CM case, cf. Example* 2.2.3. The ring $\mathrm{End}_{\overline{\mathbf{Q}}}(X_0)$ is $\mathbf{Z}$. Then $\mu^{\mathrm{disc}} = 0$ and $\mu = \mu^{\mathrm{cont}}$ is the standard Sato-Tate measure on $[-2, 2]$, namely

$$\mu = \frac{1}{2\pi}\sqrt{4 - z^2}\, dz \;=\; \frac{2}{\pi}\sin^2\!\alpha\, d\alpha, \text{ with } z = 2\cos\alpha,\ 0 \leqslant \alpha \leqslant \pi.$$

The $k$-th moment $\mu(\varphi_k)$ is 0 for $k$ odd; when $k$ is even, it is equal to the Catalan[6] number $C_{k/2} = \binom{k}{k/2} - \binom{k}{k/2-1} = \frac{1}{k/2+1}\binom{k}{k/2}$. For $k = 0, ..., 12$, this gives [1,0,1,0,2,0,5,0,14,0,42,0,132]. One has $\mu(\varphi_k) \sim \frac{4}{\sqrt{2\pi}} 2^k/k^{3/2}$ when $k \to \infty$, $k$ even.

[The proof in the non-CM case is recent. It relies on the modularity theorem for elliptic curves over $\mathbf{Q}$ originating with Wiles [Wi 95], combined with new results on the $L$-functions of symmetric powers; see [CHT 08], [Ta 08], [BLGHT 11] and the reports of Carayol [Ca 07], Clozel [Cl 08] and Harris [Ha 06], [Ha 09]. In all these proofs, the $\ell$-adic representations play an essential role.]

Note the very different behavior of the two cases near the extremities of the interval $I$ : when $\varepsilon \to 0$, the density of the $p$ with $2 - \varepsilon \leqslant f(p) \leqslant 2$ is asymptotically $\frac{1}{2\pi}\varepsilon^{1/2}$ in the CM case, and $\frac{2}{3\pi}\varepsilon^{3/2}$ in the non-CM case. For $\varepsilon = \frac{1}{100}$, this gives roughly $\frac{1}{60}$ and $\frac{1}{5000}$. This suggests that finding $p$ such that $N_X(p)$ is close to the minimum (or maximum) possible value is harder in the non-CM case than in the CM-case.

(For an explanation of the exponents 1/2 and 3/2 in terms of the dimension of the Sato-Tate group, see §8.4.4.4.)

*Exercise.* Let $e$ be an integer $> 0$. Show that the sequence $f(p^e)$ is equidistributed in $[-2, 2]$ with respect to a measure $\mu_e$.

[Hint. Use the fact that $f(p^e)$ can be written as a polynomial of degree $e$ in $f(p)$, e.g. $f(p^2) = f(p)^2 - 2$.]

When $e = 2$, show that $\mu_e = \frac{1}{2}\delta_{-2} + \frac{1}{2\pi}\frac{dz}{\sqrt{4-z^2}}$ in the CM case, and $\mu_e = \frac{dz}{2\pi}\sqrt{\frac{2-z}{2+z}}$ in the non-CM case.

---

[6]The reader shall find in [St 99, exerc. 6.19] sixty-six different combinatorial definitions of these ubiquitous numbers. See also [OEIS,A000108].

When $e > 2$, show that $\mu_e = \frac{1}{\pi}\frac{dz}{\sqrt{4-z^2}}$ in the non-CM case, while in the CM case, one has $\mu_e = \frac{1}{2}\delta_{z_0} + \frac{1}{2\pi}\frac{dz}{\sqrt{4-z^2}}$, with $z_0 = 0$ if $e$ is odd, $z_0 = 2$ if $e \equiv 0 \pmod 4$, and $z_0 = -2$ if $e \equiv 2 \pmod 4$.
[For a generalization, see the Exercise at the end of §8.4.2.]

## 8.2. The Sato-Tate correspondence

The correspondence connects :

- cohomological data (§8.2.1)

and

- Lie groups data (§8.2.2 and §8.2.3).

These data are related by several axioms, the main one being an equidistribution property relative to the Haar measure.

We shall say that a set of cohomological data *satisfies the Sato-Tate conjecture* if there are Lie groups data that are related[7] to it.

### 8.2.1. Cohomological data

The data are :
$(D_1)$ *A finite family of smooth projective varieties* $X^\lambda$.
$(D_2)$ *For every index* $\lambda$, *an integer* $w_\lambda \in \mathbf{N}$.
$(D_3)$ *A finite subset* $S$ *of* $P$, *such that every* $X^\lambda$ *has good reduction outside* $S$.

[This means that, for every $\lambda$, there exists a smooth projective scheme over $\operatorname{Spec} \mathbf{Z} - S$, whose generic fiber is the $\mathbf{Q}$-variety $X^\lambda$ [8].]

We shall be interested in the $\ell$-adic cohomology of $\overline{X}^\lambda$ with proper support in degree $w_\lambda$ ; as usual we denote it by $H^{w_\lambda}(X^\lambda, \ell)$, and we write $n_\lambda$ the corresponding Betti number, i.e. $\dim H^{w_\lambda}(X^\lambda, \ell)$.

If $p \notin S$, we denote by $P_\lambda(p,T)$ the characteristic polynomial of the (geometric) Frobenius of $p$, acting on $H^{w_\lambda}(X^\lambda, \ell)$. By Deligne's theorem (see §4.5 and §4.8.2), its coefficients belong to $\mathbf{Z}$, and its roots are $p$-Weil integers

$$P_\lambda(p,T) = T^{n_\lambda} - a_1(\lambda,p)T^{n_\lambda - 1} + a_2(\lambda,p)T^{n_\lambda - 2} + \dots ,$$

[7] Note that we are not claiming that the Lie data related to given cohomological data are unique. To have uniqueness, it seems necessary to use the theory of motives, as in [Se 94, §13]. See also the $\ell$-adic construction of §8.3 below.

[8] To be coherent with our usual conventions, the scheme over $\operatorname{Spec} \mathbf{Z} - S$ is the one that should be denoted by $X^\lambda$ ; the corresponding scheme over $\mathbf{Q}$ should then be $X_0^\lambda$. We dispense with the 0 index in order to simplify the notation.

the first coefficient $a_1(\lambda, p)$ is the trace of Frobenius, i.e. $h^{w_\lambda}_{X^\lambda}(p)$ with the notation of §7.1.3 ; we shall write it $h^\lambda(p)$.

We shall also need the *normalized* characteristic polynomial associated with $p$ and $\lambda$, i.e. the one where every root is divided by $p^{w_\lambda/2}$ :

$$P^1_\lambda(p,T) = p^{-n_\lambda w_\lambda/2} P_\lambda(p, p^{w_\lambda/2}T) = T^{n_\lambda} - \frac{a_1(\lambda,p)}{p^{w_\lambda/2}} T^{n_\lambda - 1} + \frac{a_2(\lambda,p)}{p^{w_\lambda}} T^{n_\lambda - 2} - \dots .$$

We put $f^\lambda(p) = \frac{a_1(\lambda,p)}{p^{w_\lambda/2}} = h^\lambda(p)/p^{w_\lambda/2}$ ; this is the normalized trace already defined in §8.1.1. Since the roots of $P^1_\lambda(p,T)$ have absolute value 1, the number $f^\lambda(p)$ belongs to the interval $[-n_\lambda, n_\lambda]$.

### 8.2.2. The main Lie groups data

There are three of them :

$(ST_1)$ *A compact real Lie group $K$, the "Sato-Tate group".*

$(ST_2)$ *For every $\lambda$, a continuous linear representation $r_\lambda : K \to \mathbf{GL}_{n_\lambda}(\mathbf{C})$.*

$(ST_3)$ *For every $p \in P - S$, a conjugacy class $s_p \in \mathrm{Cl}\, K$.*

For convenience, we assume that $K$ is not uselessly too large, i.e. :

$(A_0)$ *The natural map $(r_\lambda) : K \to \prod_\lambda \mathbf{GL}_{n_\lambda}(\mathbf{C})$ is injective.*
[Hence $K$ can be identified with a compact subgroup of $\prod_\lambda \mathbf{GL}_{n_\lambda}(\mathbf{C})$.]

*Remark.* It is sometimes more convenient to rewrite $(ST_2)$ with $\mathbf{GL}_{n_\lambda}(\mathbf{C})$ replaced by $\mathbf{GL}(V_\lambda)$ where $V_\lambda$ is a complex vector space of dimension $n_\lambda$. From a motivic viewpoint (as in [Se 94]), a natural choice for $V_\lambda$ is $H^{w_\lambda}(X^\lambda(\mathbf{C}), \mathbf{C})$, and the identity component of $K$ is conjecturally the so-called Mumford-Tate group (or rather a maximal compact subgroup of it).

We shall say that the data $(ST_1), (ST_2), (ST_3)$ are *related* to those of §8.2.1 if the following axioms $(A_1)$ and $(A_2)$ hold :

$(A_1)$ *For every $p \notin S$, the characteristic polynomial of $r_\lambda(s_p)$ is equal to the normalized polynomial $P^1_\lambda(p,T)$ defined above.*
[As usual, we denote by $s_p$ any element of the class $s_p$ ; hence $r_\lambda(s_p)$ is an element of $\mathbf{GL}_{n_\lambda}(\mathbf{C})$ that is defined up to conjugation, and its characteristic polynomial is uniquely defined.]
In particular, we have $\mathrm{Tr}\, r_\lambda(s_p) = f^\lambda(p)$ for every $p \notin S$ and every $\lambda$.

$(A_2)$ *The $s_p$ are equidistributed in $\mathrm{Cl}\, K$ with respect to the Haar measure.*

By "the Haar measure" of $\mathrm{Cl}\,K$, we mean the image $\mu_{\mathrm{Cl}}$ of the normalized Haar measure $\mu_K$ of $K$ by the natural projection $\pi : K \to \mathrm{Cl}\,K$. In other words, $\mu_{\mathrm{Cl}}(\varphi) = \mu_K(\varphi \circ \pi)$ for every continuous function $\varphi$ on $\mathrm{Cl}\,K$.

It is convenient to reformulate axiom $(A_2)$ in terms of characters :

$(A_2')$ *For every continuous complex character* $\psi$ *of* $K$, *one has*

$$\lim_{x\to\infty} \frac{1}{\pi_S(x)} \sum_{p\leqslant x} \psi(s_p) = <\psi,1>, \quad \textit{where} \quad <\psi,1> = \mu_K(\psi) = \int_K \psi.$$

The equivalence of $(A_2)$ and $(A_2')$ follows from the fact that the linear combinations of characters are dense in the space of all complex continuous class functions. Note that $<\psi,1>$ is the coefficient of 1 in the expansion of $\psi$ as a sum of irreducible characters; in particular, it belongs to $\mathbf{N}$.

One may reformulate $(A_2)$ and $(A_2')$ as :

$(A_2'')$ $\sum_{p\leqslant x} \psi(s_p) = o(x/\log x)$ *for every complex continuous irreducible character* $\psi$ *of* $K$, *distinct from* 1.

*Relations with L-functions and automorphic forms.*

In all the cases where $(A_2)$ has been proved, the proof relies on the properties of the $L$-function $L(s,\psi)$ associated with the character $\psi$. Such a function is defined, à la Artin (cf. [Ar 23]), by the Euler product :

$$L(s,\psi) = \prod_{p\notin S} 1/\det(1 - r_\psi(s_p)p^{-s}),$$

where $r_\psi$ is a representation of $K$ with character $\psi$. The product converges absolutely for $\mathrm{Re}(s) > 1$. The main problem is to extend it further : if possible, to a meromorphic function on $\mathbf{C}$. If one only (!) wants $(A_2)$ it is enough to have holomorphy, and non-vanishing, on the line $\mathrm{Re}(s) = 1$ whenever $\psi \neq 1$; this is a standard consequence of the Wiener-Ikehara tauberian theorem, applied to the logarithmic derivative of $L(s,\psi)$, see e.g. [Se 68, Appendix to Chap.I]. Such meromorphic continuations are usually obtained by proving that $L(s,\psi)$ coincides (except for a finite number of factors) with the $L$-function associated with an automorphic form. When that is the case, the error term $o(x/\log x)$ of $(A_2'')$ can be replaced by $O(x/(\log x)^m)$ for every $m > 0$; one expects that it is $O(x^{\frac{1}{2}}\log x)$, as in the standard GRH. From that point of view, the Sato-Tate group $K$ appears as a quotient of the would-be *automorphic Langlands group*, whose existence and structure are discussed in [La 79] and [Ar 02].

### 8.2.3. Other Lie groups data and axioms

The data and the axioms of the previous section are strong enough to imply the equidistribution properties of §8.1 (except for §8.1.3.2 and the second half of §8.1.3.3). However, both the motivic point of view and the automorphic one suggest more such properties. Here are some of them :

8.2.3.1. *The group $K/K^0$ as a Galois group.*

Let $K^0$ be the identity component of the Sato-Tate group $K$. The quotient $K/K^0$ is finite. The following axiom says that it can be viewed as a Galois group. More precisely :

$(A_3)$ *There exists a continuous surjective homomorphism $\varepsilon : \Gamma_S \to K/K^0$ such that, for every $p \notin S$, the image of $s_p$ in $\mathrm{Cl}\, K/K^0$ is equal to $\varepsilon(g_p)$, where $g_p$ is the geometric Frobenius of $p$ in $\Gamma_S$.*

Equivalently : there exists a finite Galois extension $E/\mathbf{Q}$, unramified outside $S$, with Galois group $K/K^0$, for which the geometric Frobenius are the images in $K/K^0$ of the $s_p$.

8.2.3.2. *Hodge circles.*

• *Hodge numbers.* If $p, q \in \mathbf{N}$ are such that $p + q = w_\lambda$, we denote by $h(p, q, \lambda)$ the dimension of the $(p, q)$-component of $H^\lambda$. With standard notation, we have

$$h(p, q, \lambda) = \dim H^p(X^\lambda, \Omega^q).$$

• *The Hodge circles of $K$.* Let $\mathbf{U} = \{u \in \mathbf{C}^\times \,|\, u\bar{u} = 1\}$ be the unit circle in $\mathbf{C}^\times$, and let $\theta : \mathbf{U} \to K$ be a continuous homomorphism. We shall say that $\theta$ is a *Hodge circle* if, for every $u \in \mathbf{U}$, the eigenvalues of $r_\lambda(\theta(u))$ are the $u^{p-q}$, with multiplicity $h(p, q, \lambda)$ ; equivalently, $\mathrm{Tr}\ r_\lambda(\theta(u)) = \sum_{p+q=w_\lambda} h(p, q, \lambda) u^{p-q}$ for every $u \in \mathbf{U}$. [Hence, knowing the weight $w$ and the action of $\mathbf{U}$ gives the Hodge numbers.]

We add to the list of data of §8.2.2 the following :

$(ST_4)$ *A homomorphism $\theta : \mathbf{U} \to K$ that is a Hodge circle.*

*Remark.* As explained in §8.2.2, we could have defined $r_\lambda$ as giving the action of $K$ on $V_\lambda = H^{w_\lambda}(X^\lambda(\mathbf{C}), \mathbf{C})$. If so, we could restate $(ST_4)$ in a more precise form, as follows : define a homomorphism $\theta : \mathbf{U} \to \prod_\lambda \mathbf{GL}(V_\lambda)$ by the condition that every $u \in \mathbf{U}$ acts on the $(p, q)$-component of $V_\lambda$ by the homothety $u^{p-q}$ ; then, the refined form of $(ST_4)$ is the requirement that *the image of $\theta$ is contained in $K$*. This is the starting point of the Mumford-Tate construction of $K^0$, cf. [Mu 66].

8.2.3.3. *The central element $\omega$.*

Let us define $\omega \in K$ by the formula $\omega = \theta(-1)$, where $\theta$ is as in $(ST_4)$. We have $\omega^2 = 1$. Moreover :

**Proposition 8.1.** *The image of $\omega$ by $r_\lambda : K \to \mathbf{GL}_{n_\lambda}(\mathbf{C})$ is $(-1)^{w_\lambda}$.*

*Proof.* It follows from the definition of $h$ that the eigenvalues of $r_\lambda(\omega)$ are of the form $(-1)^{p-q}$, with $p+q = w_\lambda$ ; since $(-1)^{p-q} = (-1)^{p+q} = (-1)^{w_\lambda}$, the proposition follows.

**Corollary 8.2.** *The element $\omega$ belongs to the center of $K$; it does not depend on the choice of a Hodge circle.*

*Proof.* The proposition shows that the image of $\omega$ in $\prod_\lambda \mathbf{GL}_{n_\lambda}(\mathbf{C})$ belongs to the center of that group. By $(A_0)$, this implies that $\omega$ is central of $K$. A similar argument shows that $\omega$ does not depend on the choice of $h$.

8.2.3.4. *The element $\gamma$.*

We add to the data :

$(ST_5)$ *An element $\gamma$ of $K$, with $\gamma^2 = \omega$, such that $\gamma\theta(u)\gamma^{-1} = \theta(u^{-1})$ for every $u \in \mathbf{U}$.*

[Loosely speaking, this means that $\gamma$ interchanges the $(p, q)$ and $(q, p)$ components of the Hodge decompositions. Note that this amounts to a homomorphism of the "Weil group of $\mathbf{R}$" into $K$, cf. [Ta 79, 1.4.3].]

The element $\gamma$ should satisfy the following condition :

$(A_4)$ i) *The image of $\gamma$ in $K/K^0 =$* Gal $E/\mathbf{Q}$ (cf. §8.2.3.1) *belongs to the conjugacy class of the complex conjugation.*

ii) *If $w_\lambda$ is odd, the trace of $r_\lambda(\gamma)$ is* 0.

iii) *If $w_\lambda$ is even, the characteristic polynomial of $(-1)^{w_\lambda/2} r_\lambda(\gamma)$ is the same as the characteristic polynomial of the complex conjugation, acting on $H^\lambda(X^\lambda(\mathbf{C}), \mathbf{Q})$.*

8.2.3.5. *Invariant bilinear forms.*

It follows from $(A_1)$ that the traces of the $r_\lambda(s_p)$ are real numbers, and because of the density property $(A_2)$ the same is true for all the $r_\lambda(z)$ with $z \in K$. This is equivalent to saying that *the representation $r_\lambda$ is isomorphic to its dual.* We ask more :

$(A_5)$ *If $w_\lambda$ is even (resp. odd), the representation $r_\lambda$ leaves invariant a non-degenerate symmetric (resp. alternating) bilinear form.*

8.2.3.6. *Other properties of the Sato-Tate correspondence.*

We mention a few more, without giving details :

i) The smallest closed subgroup of $K^0$ containing all Hodge circles should be $K^0$ itself.

ii) The "bad primes" (i.e. those belonging to $S$) should define subgroups of $K$ involving the ramification (through the so-called local Weil group, cf. [Ta 79, §1.4.1]). From that point of view, the Hodge circle comes from the Weil group over $\mathbf{C}$ and its complement $\gamma$ of 8.2.3.4 comes from the Weil group over $\mathbf{R}$.

In particular, every bad prime $p$ should determine a unipotent conjugacy class $u_p$ in the complexification of $K$. The class $u_p$ should be trivial if and only if there is "potential good reduction" at $p$, in the sense of [Se 94, end of §12]. When $u_p \neq 1$, one may interpret it (via Jacobson-Morozov) as an $A_1$-type subgroup of $K^0$.

iii) The tensor invariants of $K$ should be related to the semi-invariants (with respect to Tate twists) of the $\ell$-adic representations; axiom $(A_5)$ is a typical example.

## 8.3. An $\ell$-adic construction of the Sato-Tate group

There are two different methods for constructing Lie data, in the sense of §8.2.2, that (hopefully) correspond to a given set of cohomological data, in the sense of §8.2.2 :

- the motivic method;
- the $\ell$-adic method.

The first one is the most natural, but it has the defect of depending on the so-called standard conjectures of motivic theory : Hodge, Hodge-Tate, positivity, etc. The second one is more direct, but depends on two auxiliary choices : a prime number $\ell$ and an embedding $\iota : \overline{\mathbf{Q}}_\ell \to \mathbf{C}$. Neither of the two gives any clue on how to prove the basic equidistribution axiom $(A_2)$.

Since the motivic approach is given in [Se 94, §13], I shall only describe here the $\ell$-adic one.

### 8.3.1. Notation

Let $\{X^\lambda, w_\lambda, S\}$ be cohomological data, and let $\ell$ be a prime number; put $S_\ell = S \cup \{\ell\}$. By enlarging the family $X^\lambda$ (which only makes the problem more difficult), we may assume that one of the $X^\lambda$, say $X^1$, is the projective line $\mathbf{P}_1$ and that $w_1 = 2$; this insures that the corresponding Galois representation is given by $\chi_\ell^{-1}$, where $\chi_\ell$ is the $\ell$-th cyclotomic character.

Let $E_\lambda = H^{w_\lambda}(X^\lambda, \ell)$. Let $E = \bigoplus_\lambda E_\lambda$; it is a finite dimensional $\mathbf{Q}_\ell$-vector space, on which the Galois group $\Gamma_{S_\ell}$ acts.

### 8.3.2. The $\ell$-adic groups

The action of $\Gamma_{S_\ell}$ on the $E_\lambda$ and on their direct sum $E$ defines a homomorphism

$$\rho_\ell = (\rho_{\ell,\lambda}) : \Gamma_{S_\ell} \to \prod \mathbf{GL}(E_\lambda) \subset \mathbf{GL}(E).$$

Let $G_\ell$ be the image of $\rho_\ell$ and let $G_\ell^{\mathrm{zar}}$ be its Zariski closure, which we view as a $\mathbf{Q}_\ell$-algebraic subgroup of $\mathbf{GL}_n$, where $n = \dim E = \sum n_\lambda$. It follows from a theorem of Bogomolov ([Bo 80]) that $G_\ell$ is open[9] in the group $G_\ell^{\mathrm{zar}}(\mathbf{Q}_\ell)$ of the $\mathbf{Q}_\ell$-points of $G_\ell^{\mathrm{zar}}$.

Let us denote by $N : G_\ell^{\mathrm{zar}} \to \mathbf{G}_m$ the $\lambda$-component of $G_\ell^{\mathrm{zar}} \to \mathbf{GL}_n$ relative to $\lambda = 1$. If $p \notin S_\ell$, the image by $N$ of the geometric Frobenius $g_p$ is equal to $p$. The kernel of $N$ will be denoted by $G_\ell^{1,\mathrm{zar}}$.

We shall also use the homomorphism $\boldsymbol{w} : \mathbf{G}_m \to G_\ell^{\mathrm{zar}}$ defined in the following way : if $x$ is any point of $\mathbf{G}_m, \boldsymbol{w}(x)$ acts on each $E_\lambda$ by the homothety $x^{w_\lambda}$.

[The fact that the image of $\boldsymbol{w}$ is contained in $G_\ell^{\mathrm{zar}}$ is proved by a nice argument, due to Deligne : one selects a prime $p \notin S_\ell$, and one shows (by using the Weil-Deligne estimates) that Im $\boldsymbol{w}$ is contained in the minimal algebraic subgroup of $\mathbf{GL}_n$ containing $g_p$.]

The image of $\boldsymbol{w}$ is contained in the center of $G_\ell^{\mathrm{zar}}$. Moreover, the composite map $\mathbf{G}_m \overset{\boldsymbol{w}}{\to} G_\ell^{\mathrm{zar}} \overset{N}{\to} \mathbf{G}_m$ is equal to 2 in $\mathrm{End}(\mathbf{G}_m) = \mathbf{Z}$.

### 8.3.3. Definition of the Lie groups data

We follow the same method as in §5.3.1, i.e. we choose an embedding $\iota : \overline{\mathbf{Q}}_\ell \to \mathbf{C}$. By the base change $\iota$, the $\mathbf{Q}_\ell$-algebraic groups $G_\ell^{\mathrm{zar}}$ and $G_\ell^{1,\mathrm{zar}}$ give $\mathbf{C}$-algebraic groups $G_{\ell,\iota}^{\mathrm{zar}}$ and $G_{\ell,\iota}^{1,\mathrm{zar}}$. Let us denote simply by $G$ and $G^1$ the groups of their $\mathbf{C}$-points, i.e. $G = G_{\ell,\iota}^{\mathrm{zar}}(\mathbf{C})$ and $G^1 = G_{\ell,\iota}^{1,\mathrm{zar}}(\mathbf{C})$. We have an exact sequence

$$1 \to G^1 \to G \overset{N}{\to} \mathbf{C}^\times \to 1,$$

and the composite map $\mathbf{C}^\times \overset{\boldsymbol{w}}{\to} G \overset{N}{\to} \mathbf{C}^\times$ is $u \mapsto u^2$.

We can now define the required Lie groups data $(K, r_\lambda, s_p)$.

First *the Sato-Tate group* $K$ : it is a maximal compact subgroup of the complex Lie group $G^1$. [Note that two such subgroups are conjugate; hence it does not matter which one we choose.]

---

[9]Bogomolov's theorem requires that $E$ has a Hodge-Tate decomposition at $\ell$; the existence of such a decomposition has been proved by Fontaine-Messing [FM 85] under the hypotheses $\ell \notin S$ and $\ell > \sup w_\lambda$, and then by Faltings [Fa 88] in the general case.

Next *the representations* $r_\lambda : K \to \mathbf{GL}_{n_\lambda}(\mathbf{C})$ : they are the restrictions to $K$ of the linear representations of $G$ obtained by base change from the $(\rho_{\ell,\lambda})$.

As for *the conjugacy classes* $s_p$, $p \notin S_\ell$, they are defined as follows : the element $g_p$ of $G_\ell^{\mathrm{zar}}(\mathbf{Q}_\ell)$ gives, by the base change $\iota$, an element of $G$ that we denote by $g_{p,\iota}$ ; it is well defined up to conjugation, and its image by $N : G \to \mathbf{C}^\times$ is equal to $p$. Hence the element $g'_p = w(p^{-\frac{1}{2}})g_{p,\iota}$ belongs to $G^1$. [Note that $p^{\frac{1}{2}}$ is unambiguous in $\mathbf{C}$; it would not be in $\overline{\mathbf{Q}}_\ell$.] The eigenvalues of $g'_p$ in the representations $r_\lambda$ have absolute value 1. Let $g_p^{\mathrm{ss}}$ be the semisimple component of $g'_p$, with respect to the Jordan decomposition. Since $g_p^{\mathrm{ss}}$ is semisimple, and all its eigenvalues have absolute value 1, it is contained in a compact subgroup of $G^1$, and hence it is conjugate to an element $s_p$ of $K$. This is the element we wanted to define; note that it is unique, up to conjugation, because the natural maps $\mathrm{Cl}\,K \to \mathrm{Cl}\,G^1 \to \mathrm{Cl}\,G$ are injective[10].

Axioms $(A_0)$ and $(A_1)$ are satisfied (with $S$ replaced by $S_\ell$ in case $\ell \notin S$). As for $(A_2)$, it is a reasonable conjecture that it is satisfied for every $\ell$.

*Remark.* It is also conjectured that the groups $G_\ell^{\mathrm{zar}}, G, G^1, \ldots$ are reductive groups, and that the Frobenius elements $g_p$ (and hence $g'_p$) are semisimple.

### 8.3.4. Complements

Most of the data and axioms of §8.2.3 can be defined and checked. Namely :

*The group $K/K^0$ as a Galois group, cf. §8.2.3.1.* Let us write, in Topology style, $\pi_0(K)$ for $K/K^0$ and similarly for other topological or algebraic groups. We have a natural homomorphism

$$\Gamma_{S_\ell} \to \pi_0(G_\ell^{\mathrm{zar}}) \simeq \pi_0(G)$$

that is surjective with open kernel. We also have homomorphisms

$$\pi_0(K) \to \pi_0(G^1) \to \pi_0(G).$$

The left one is an isomorphism (by a general property of maximal compact subgroups). The right one is also an isomorphism (because the extension $1 \to G^1 \to G \to \mathbf{C}^\times \to 1$ splits, as follows from the existence of Hodge circles, see below). By composing these maps, we get a continuous surjective

[10] The injectivity of $\mathrm{Cl}\,K \to \mathrm{Cl}\,G^1$ is proved by reduction to the case where $G^1$ is reductive, in which case it follows from standard properties of the Cartan decomposition $G^1 = K.P$, where $P = \exp(i.\mathrm{Lie}\,K)$. The injectivity of $\mathrm{Cl}\,G^1 \to \mathrm{Cl}\,G$ is clear since $G$ is generated by $G^1$ and the central subgroup $w(\mathbf{C}^\times)$.

homomorphism $\varepsilon : \Gamma_{S_\ell} \to K/K^0$, as desired. Axiom $(A_3)$ follows from the construction. [The Galois extension of $\mathbf{Q}$ associated with the kernel of $\varepsilon$ *is independent of the choice of* $\ell$; this can be proved by either one of the two methods given in [Se 91].]

*Hodge circles, cf. §8.3.2.* By using Hodge-Tate decompositions at $\ell$ one can define (using, e.g., [Se 79]) a homomorphism $\boldsymbol{h} : \mathbf{C}^\times \to G$ that gives the $p$-part of the Hodge decomposition, in the following sense : for every $u \in \mathbf{C}^\times$, and every $p \in \mathbf{N}$, $u^p$ is an eigenvalue of $r_\lambda(\boldsymbol{h}(u))$ with multiplicity $h(p, w_\lambda - p)$. The map $\mathbf{U} \to G^1$ defined by $u \mapsto \boldsymbol{h}(u^2)\boldsymbol{w}(u^{-1})$ is conjugate in $G^1$ to a homomorphism $\theta : \mathbf{U} \to K$ which is a Hodge circle.

*The elements $\omega$ and $\gamma$, cf. §8.2.3.3 and §8.2.3.4.* The element $\omega$ is defined as $\theta(-1) = \boldsymbol{w}(-1)$. To define $\gamma$, consider first the complex conjugation $c$, viewed as an element of $G_\ell$, hence also of $G$. Since $\chi_\ell(c) = -1$, we have $N(c) = -1$. Since $N(\boldsymbol{w}(i)) = i^2 = -1$, the element $\boldsymbol{w}(i).c$ of $G$ belongs to $G^1$, and its square is equal to $\omega$. It is thus conjugate to an element $\gamma$ of $K$ that has all the properties required in $(A_4)$.

[On the other hand, I do not see how to prove without using motivic theory that $\gamma$ has the property stated in $(ST_5)$, i.e. that it can be chosen in such a way that it "inverts" a Hodge circle.]

*Bilinear forms, cf. §8.2.3.5.* The existence of these forms follows from the corresponding fact for the $\ell$-adic representations given by cohomology, which is itself a consequence of Hodge theory.

## 8.4. Consequences of the Sato-Tate conjecture

Our aim now is to show that "8.2 $\Rightarrow$ 8.1", i.e. that the statements of §8.1 are consequences of the Sato-Tate conjecture, as formulated in §8.2.

### 8.4.1. The theorem

Let $X, w, h_X^w(p), f_X(p) = h_X^w(p)/p^{w/2}, S, \ldots$ be as in §8.1.1. By Proposition 7.7, we may find smooth projective varieties $X^\lambda$, and integers $m_\lambda \in \{\pm 1\}$, such that

$$h_X^w = \sum_\lambda m_\lambda h_{X^\lambda}^w \quad \text{and hence} \quad f_X = \sum_\lambda m_\lambda f_{X^\lambda}.$$

We also assume that $S$ is large enough so that all the $X_\lambda$ have good reduction outside $S$. If we put $w_\lambda = w$ for every $\lambda$, the data $\{X^\lambda, w_\lambda, S\}$ are "cohomological data" in the sense of §8.2.1.

**Theorem 8.3.** *Assume that $\{X^\lambda, w_\lambda, S\}$ satisfy the Sato-Tate conjecture, i.e. are related to Lie groups data $\{K, r_\lambda, s_p\}$ as in* §8.2.3 *and* §8.2.4. *Then* *the equidistribution statements on $f_X$ made in* §8.1.2, §8.1.3 *and* §8.1.4 *are true.*

*Remark.* The proof will not use axioms $(A_4)$ and $(A_5)$.

### 8.4.2. Proof of Theorem 8.3 : supports and moments

Let $\psi_\lambda = \mathrm{Tr}\ r_\lambda$ be the character of the representation $r_\lambda$ of $K$. Define a virtual character $\psi$ of $K$ by the formula $\psi = \sum_\lambda m_\lambda \psi_\lambda$, where the $m_\lambda$ are as above. The values of $\psi$ are real; choose a closed interval $I = [-C, C]$ containing them. For every $p \notin S$, we have $\psi(s_p) = \sum m_\lambda \psi_\lambda(s_p)$; hence, by applying $(A_1)$, we get

$$\psi(s_p) = \sum m_\lambda f_{X^\lambda}(p) = f_X(p).$$

8.4.2.1. *Definition of* $\mu$. Let $\mu_K$ be the normalized Haar measure of $K$, and let $\mu = \psi_* \mu_K$ be the image of $\mu_K$ by the continuous map $\psi : K \to I$. It is a positive measure of mass 1 on $I$. We may also view it as the image of the measure $\mu_{\mathrm{Cl}}$ by the map $\mathrm{Cl}\, K \to I$ defined by $\psi$. By $(A_2)$, the $s_p$ are $\mu_{\mathrm{Cl}}$-equidistributed. Hence their images $f_X(p) = \psi(s_p)$ are $\mu$-equidistributed. This proves the equidistribution statement of §8.1.2.

8.4.2.2. *Proof of 8.1.3.2.* Assume that $w$ is odd. Then the central element $\omega$ defined in §8.2.3.3 is such that $\psi(\omega x) = -\ \psi(x)$ for every $x \in K$, cf. Proposition 8.1. Since $x \mapsto \omega x$ leaves invariant the Haar measure $\mu_K$, we have a similar result for its image by $\psi$ : the measure $\mu$ is invariant by $z \mapsto -z$. This proves 8.1.3.2; it follows from it that the $k$-th moments of $\mu$ are 0 when $k$ is odd, as asserted in §8.1.3.4.

8.4.2.3. *Proof of 8.1.3.3.* The support of $\mu$ is the image by $\psi$ of the support of $\mu_K$, hence it is equal to $\psi(K)$; since the $f_X(p)$'s are $\mu$-equidistributed, they are dense in the support of $\mu$. If moreover $X$ is projective and smooth, we can take a family $(X^\lambda)$ consisting of $X$ alone, and choose for $C$ the $w$-th Betti number of $X$, which is equal to $f_X(1)$. Suppose now that $w$ is odd. The image by $\psi$ of $\omega$ is $-C$. Let $J$ be the image of a Hodge circle by $\psi$ (more correctly : the image of $\psi \circ \theta : \mathbf{U} \to I$). It is a connected subset of $[-C, C]$ that contains both $C$ and $-C$; hence its is equal to $[-C, C]$. This shows that $\psi(K) = [-C, C]$.

8.4.2.4. *Proof of 8.1.3.4.* The $k$-th moment $\mu(\varphi_k)$ of $\mu$ is the integral $\int_I z^k \mu(z) = \int_K \psi^k(x) \mu_K(x)$, which can also be rewritten as the scalar product $< \psi^k, 1 >$. It is the coefficient of 1 in the expansion of the virtual character $\psi^k$ as a linear combination of irreducible characters. Hence it belongs to $\mathbf{Z}$, and, if $\psi$ is effective, it is $\geqslant 0$. (Moreover, if $\psi \neq 0$, one has $\int \psi^k \mu_K \geqslant 1$ for every even $k$ : all the even moments are $> 0$.)

*Exercise.* Let $e$ be an integer. Show that the sequence $p \mapsto f_X(p^e)$ is $\mu_e$-equidistributed, where $\mu_e$ is a measure on $I$, with $\mathrm{Supp}(\mu_e) \subset \mathrm{Supp}(\mu)$, having the following properties :

a) There is a decomposition of $\mu_e$ as $\mu_e^{\mathrm{disc}} + \mu_e^{\mathrm{cont}}$ with properties similar to those of §8.1.3.1.

b) For every $k \geqslant 0$, the $k$-th moment of $\mu_e$ is an integer.

c) If $k, w$ and $e$ are odd, $\mu_e$ is invariant by $z \mapsto -z$.

[Hint. Define $\mu_e$ as the image of $\mu_K$ by the virtual character $\Psi^e\psi : K \to I$.]

d) Assume that $K$ is connected, and let $h$ be the upper bound of the Coxeter numbers of its simple components (with the convention that $h = 1$ if $K$ is a torus). Show that, if $|e| > h$, all the $\mu_e$'s are the same.

[Hint. Let $m_e : K \to K$ be the map $x \mapsto x^e$, and let $\mu_{K,e}$ be the image of $\mu_K$ by $m_e$ ; view $\mu_{K,e}$ as a measure on $\mathrm{Cl}\, K$. By a theorem of E.M. Rains [Ra 03, Th.4.1], if $|e| > h$ (resp. if $|e| \geqslant h$ if the center of $K$ is connected), $\mu_{K,e}$ is independent of $e$; more precisely, if $T$ is a maximal torus of $K$, $\mu_{K,e}$ is equal to the image by $T \to K \to \mathrm{Cl}\, K$ of the normalized Haar measure of $T$. Use this fact to prove d).]

### 8.4.3. Proof of Theorem 8.3 : structure of $\mu$

8.4.3.1. *Proof of 8.1.3.1 - Definition of the decomposition of $\mu$.* If $\sigma$ is an element of the finite group $K/K^0$, let $K_\sigma$ be the corresponding $K^0$-coset of $K$; it is a connected compact manifold. Let $\mu_{K,\sigma}$ be the restriction to $K_\sigma$ of the Haar measure $\mu_K$ ; it is a positive measure. The measure $\mu = \psi_*\mu$ is the sum of the measures $\mu_\sigma = \psi_*\mu_{K,\sigma}$. Each of these measures has a mass equal to $1/N$, with $N = (K : K^0)$. Let $D$ be the subset of $K/K^0$ made up of the $\sigma$ such that $\psi$ is constant on $K_\sigma$. Define $\mu^{\mathrm{disc}}$ and $\mu^{\mathrm{cont}}$ by the formulae

$$\mu^{\mathrm{disc}} = \sum_{\sigma \in D} \mu_\sigma \quad \text{and} \quad \mu^{\mathrm{cont}} = \sum_{\sigma \notin D} \mu_\sigma .$$

We have $\mu = \mu^{\mathrm{disc}} + \mu^{\mathrm{cont}}$. Let us show that this decomposition has the properties stated in §8.1.3.1.

8.4.3.2. *Proof of 8.1.3.1 - Properties of $\mu^{\mathrm{disc}}$.* If $\sigma$ belongs to $D$, let $n_\sigma$ be the constant value of $\psi$ on $K_\sigma$. The measure $\mu_\sigma$ is obviously $\frac{1}{N}\delta_{n_\sigma}$. This shows that $\mu^{\mathrm{disc}}$ is a discrete measure - more precisely, it is a finite linear combination of Dirac measures, the coefficients being integral multiples of $\frac{1}{N}$.

It remains to prove :

**Lemma 8.4.** *If $\sigma \in D$, then $n_\sigma$ is an integer ; if moreover the weight $w$ is odd, then $n_\sigma = 0$.*

Let $\mathrm{Cl}_\sigma$ be the image of $K_\sigma$ in $\mathrm{Cl}\, K$. It is a non-empty open and closed subset of $\mathrm{Cl}\, K$. The equidistribution axiom $(A_2)$ implies that there exist two distinct primes $p$ and $p'$ with $s_p, s_{p'} \in \mathrm{Cl}_\sigma$. We have $n_\sigma = a/p^{w/2} = a'/p'^{w/2}$, where $a$ and $a'$ denote the integers $f_X(p)$ and $f_X(p')$. Assume $n_\sigma \neq 0$. If $w$ is odd, the equation $(p/p')^{w/2} = a/a'$ implies that $(p/p')^{\frac{1}{2}}$ is rational, which is impossible. If $w$ is even, the same equation shows that $a$ is divisible by $p$, hence $n_\sigma$ is an integer.

[Alternative proof for the fact that $w$ odd $\Rightarrow n_\sigma = 0$ : since the central element $\omega$ of §8.2.3.3 belongs to the image of a Hodge circle, it is contained in $K^0$, and we have $\omega.K_\sigma = K_\sigma$. If $x \in K_\sigma$, this shows that $n_\sigma = \psi(x) = \psi(\omega x)$; since $w$ is odd, $\psi(\omega x)$ is equal to $-\psi(x)$, cf. 8.4.2.2. Hence $n_\sigma = -n_\sigma$.]

8.4.3.3. *Proof of 8.1.3.1 - Properties of* $\mu^{\mathrm{cont}}$. *A lemma.* Suppose now that $\sigma \notin D$, and let us show that the measure $\mu_\sigma$ has the properties listed in §8.1.3.1. By definition, $\mu_\sigma$ is the image by $\psi : K_\sigma \to \mathbf{R}$ of the measure $\mu_{K,\sigma}$. The properties of $\mu_\sigma$ stated in §8.1.3.1 are consequences of the following lemma from differential geometry :

**Lemma 8.5.** *Let $\psi : Y \to \mathbf{R}$ be a $C^r$-function on a compact differential oriented real $C^r$-manifold $Y$ of dimension $N$(with $r \geqslant 1$). Let $C$ be the critical set of $\psi$, i.e. the set of points $y \in Y$ where $d\psi = 0$, and let $V = \psi(C)$ be the set of critical values of $\psi$. Assume that $V$ is finite. Let $\alpha$ be a $C^r$-differential form of degree $N$ over $Y$, that we view as a measure on $Y$. Assume that this measure is positive and also that $C$ has measure* 0 *for it.*

*Then the image $\psi_*\alpha$ of the measure $\alpha$ by $\psi$ has a density: it is equal to $F(z)dz$, where $F$ is a positive integrable $C^r$-function on $\mathbf{R} - V$.*

The hypotheses of the lemma are fulfilled by $Y = K_\sigma$. Indeed, the Lie group $K$ has a natural real analytic structure. Since $\psi$ is non constant on $Y$, the critical set $C$ has an empty interior; since it is analytic, its measure for $\alpha$ is 0. The function $\psi$ is constant on every connected component of $C$; hence its image $V$ in $\mathbf{R}$ is finite.

8.4.3.4. *Proof of Lemma 8.5.* Let $Y_0 = Y - \psi^{-1}(V)$, and let $\psi_0$ be the restriction of $\psi$ to $Y_0$. The map $\psi_0 : Y_0 \to \mathbf{R} - V$ is proper and smooth (in the $C^\infty$-category). We may apply to it the standard "integration over the fiber" process (see e.g. [De 09, I.2.15]). Let us briefly recall how this works :

Let $z$ be a point of $\mathbf{R} - V$, and let $Y_z$ be its inverse image in $Y_0$. If $y$ is a point of $Y_z$, we may choose a local system of coordinates $(t_1, ..., t_N)$ at $y$ such that the last coordinate $t_N$ is equal to $\psi - z$. In a neighborhood of $y$, we may write $\alpha$ as $a(t_1, ..., t_N)dt_1 \wedge \cdots \wedge dt_N$, where $a$ is $C^\infty$. On $Y_z$ the $(N-1)$-th differential form $\alpha_z = a(t_1, ..., t_{N-1}, 0)dt_1 \wedge \cdots \wedge dt_{N-1}$ is independent of the choices we have made (a suggestive notation for it is

" $\alpha/d\psi$ "). The chosen orientation of $Y$ gives an orientation of $Y_z$. Hence the measure $\alpha_z$ on $Y_z$ is well defined. We have

$$(*) \quad \alpha = \int_{z\in\mathbf{R}-V} \alpha_z dz,$$

which means that, if $\varphi$ is any continuous function with compact support on $Y_0$, then

$$(**) \quad \int_{y\in Y_0} \varphi(y)\alpha(y) = \int_{z\in\mathbf{R}-V} \Big(\int_{y\in Y_z} \varphi(y)\alpha_z(y)\Big)dz.$$

(The proof is a local computation, based on the Lebesgue-Fubini theorem for continuous functions.) This is a special case of the "integration of measures" process, see [INT, Chap.V].

Let us now define a function $F$ on $\mathbf{R}-V$ by $F(z) = \int_{Y_z} \alpha_z$ (" volume of the fiber $Y_z$ "). This is a $C^r$ function on $\mathbf{R}-V$ (differentiability of integrals depending on parameters). If we apply (**) to a function $\varphi(z)$ depending only on $z = \psi(y)$, we get

$$\int_{y\in Y_0} \varphi(y)\alpha(y) = \int_{z\in\mathbf{R}-V} \varphi(z)F(z)dz.$$

This can be restated as $(\psi_0)_*\alpha_0 = F(z)dz$, where $\alpha_0$ is the restriction to $Y_0$ of the measure $\alpha$. The function $F$ is positive, and integrable, since its integral is equal to the mass of $\alpha_0$ ; it is 0 outside $\psi(Y)$. We may thus view $F(z)dz$ as a measure on $\psi(Y)$. Let $\beta = \psi_*\alpha - F(z)dz$ ; it is a measure on a compact subset of $\mathbf{R}$. We have just shown that $\beta$ is 0 on $\mathbf{R}-V$ ; on the other hand, $V$ has measure 0 for both $F(z)dz$ and $\psi_*\alpha$, hence also for $\beta$. This shows that $\beta = 0$, i.e. $\psi_*\alpha = F(z)dz$, as wanted.

### 8.4.4. End of proof of Theorem 8.3 : density properties

8.4.4.1. If $A$ is any subset of $I$, let $P_A$ be the set of $p \notin S$ with $f_X(p) \in A$. The equidistribution of the $f_X(p)$'s imply the following statements, which generalize those of Theorem 6.8 :

- If $A$ is open, then lower-dens$(P_A) \geqslant \mu(A)$.
- If $A$ is closed, then upper-dens$(P_A) \leqslant \mu(A)$.

[*Proof of the first statement.* By definition, $\mu(A)$ is the upper bound of $\mu(\varphi)$ for all positive continuous $\varphi \leqslant 1$ that vanish outside $A$; apply the definition of equidistribution to such $\varphi$'s.]

8.4.4.2. *Proof of 8.1.4.1 : the case of a finite subset.* Suppose that $A$ is a finite subset of $I$. Let $K_A$ be the subspace of $K$ made up of the elements

$x$ such that $\psi(x) \in A$; we have $p \in P_A \iff s_p \in K_A$ (as usual, we are identifying $s_p$ with an element of $K$ defined up to conjugation). Define $D$, $N$ and $n_\sigma(\sigma \in D)$ as in 8.4.3.1 and 8.4.3.2. We may split $K$ as $K = K' \sqcup K''$, where $K' = \bigsqcup_{\sigma \in D} K_\sigma$ and $K'' = \bigsqcup_{\sigma \notin D} K_\sigma$. By intersecting with $K_A$ we have a corresponding decomposition of $K_A$ as $K_A = K'_A \sqcup K''_A$, and similarly $P_A = P'_A \sqcup P''_A$.

The set $K'_A$ is the union of the $K^0$-cosets of $K$ corresponding to the elements $\sigma \in D$ with $n_\sigma \in A$. It is open and closed in $K$, and it is stable under conjugation. It follows from the equidistribution of the $s_p$ that $P'_A$ has a density equal to $\mu_K(K'_A)$, which is equal to $\mu^{\mathrm{disc}}(A) = \mu(A)$; moreover, by $(A_3)$, $P'_A$ is an $S$-frobenian set. The set $K''_A$ is analytic with empty interior. Hence its measure is 0 (cf. Proposition 5.9), and it follows that $\mathrm{dens}(P''_A) = 0$. We thus have $\mathrm{dens}(P_A) = \mathrm{dens}(P'_A) = \mu(A)$, as asserted in 8.1.4.1.

8.4.4.3. *Proof of 8.1.4.2 : the case of an open interval.* Let us now assume that $A$ is an open interval (denoted by $J$ in 8.1.4.2) ; let $\overline{A}$ be its closure, and $\partial A = \overline{A} - A$ its boundary. We have

$$\begin{aligned} \text{lower-dens}(P_A) &\geqslant \mu(A) \quad \text{(by 8.4.4.1)}, \\ \text{upper-dens}(P_{\overline{A}}) &\leqslant \mu(\overline{A}) \quad \text{(by 8.4.4.1)}, \\ \mathrm{dens}(P_{\partial A}) &= \mu(\partial A) \quad \text{(by 8.4.4.2)}. \end{aligned}$$

Since uppe-dens$(P_{\overline{A}})$ = upper-dens$(P_A)$+dens$(P_{\partial A})$, this gives

$$\text{upper-dens}(P_A) \leqslant \mu(A),$$

and hence $\mathrm{dens}(P_A) = \mu(A)$, as asserted in 8.1.4.2.

8.4.4.4. *Moment estimates.* One can use the method of §8.4.3 to obtain asymptotic properties of the moments $\mu(\varphi_k)$ for $k \to \infty$. More precisely, assume that the character $\psi$ is effective and $\neq 0$; let $n = \psi(1)$ be its degree; assume also that the corresponding representation of $K$ is faithful, so that we may identify $K$ with a compact subgroup of $\mathbf{GL}_n(\mathbf{C})$. All the values of $\psi$ are real, and lie in the interval $I = [-n, n]$; if $x \in K$, we have $\psi(x) = n$ if and only if $x = 1$, and $\psi(x) = -n$ is possible only if $-1 \in K$ and $x = -1$. At the point 1, the second derivative of $\psi$ is the quadratic form $q : \mathfrak{g} \to \mathbf{R}$ defined by $q(\xi) = \mathrm{Tr}(\xi^2)$, where $\mathfrak{g}$ is the Lie algebra of $K$; by looking at the eigenvalues of $\xi$, one checks that $q(\xi) < 0$ for every non-zero element $\xi$ of $\mathfrak{g}$. Hence $\psi$ is a *Morse function* at the critical point 1 of $K$, and one can choose local coordinates $(x_1, ..., x_d)$ at 1 such that $\psi(x) = -\sum x_i^2$. Standard arguments then show that there exists $c > 0$ such that

$$\mu([n, n - \varepsilon]) \sim c\,\varepsilon^{d/2} \text{ when } \varepsilon > 0,\ \varepsilon \to 0, \text{ where } d = \dim K,$$

and

$$(\text{density of } \mu \text{ at } n - \varepsilon) \quad \sim c' \, \varepsilon^{d/2-1}, \text{ with } c' = cd/2.$$

Moreover, when $k$ is large, and even, the size of the $k$-th moment $\mu(\varphi_k) = \int_K \psi^k \mu_K$ only depends on the behaviour of $\psi$ near $x = 1$ and $x = -1$. Using the estimates above, one then sees that there exists $c'' > 0$ such that

$$\mu(\varphi_k) \sim c'' n^k / k^{d/2}, \text{ when } k \to \infty, \quad k \text{ even}^{11}.$$

The cases of §8.1.5.1 and §8.1.5.2 correspond to $(n, d) = (2, 1)$ and $(n, d) = (2, 3)$ respectively.

*Exercises.*

1) Give an alternative proof of 8.1.4.1 and 8.1.4.2 as follows :

Let $A$ be the union of a finite number of points and intervals in $I$, and let $K_A$ be the subset of $K$ made up of the elements $x$ with $\Phi(x) \in A$, where $\Phi$ is a given real analytic function on $K$. Show that the boundary of $K_A$ is a closed analytic subspace of $K$ with empty interior, hence has measure 0 for $\mu_K$. Conclude that $K_A$ is $\mu_K$-quarrable. When $\Phi = \psi$, this implies that $\mathrm{dens}(P_A) = \mu(A)$.

2) Let $G$ be a compact group and let $\chi$ be a virtual complex continuous character of $G$. Assume that $\chi \neq 0$.

a) Show that there exists $g \in G$ with $|\chi(g)| \geqslant 1$.

[Hint. Use the fact that $\int_G |\chi|^2 \geqslant 1$.]

b) Assume that $\chi$ is real-valued and that the integer $\chi(1)$ is even. Show that there exists $g \in G$ with $|\chi(g)| \geqslant 2$.

[Hint. There exists a commutative closed subgroup $A$ of $G$ such that the restriction of $\chi$ to $A$ is $\neq 0$. By replacing $G$ by $A$ we may assume that $G$ is commutative and write $\chi$ as $\sum n(\psi)\psi$, where the characters $\psi$ are pairwise distinct 1-dimensional, and the coefficients $n(\psi)$ are non-zero. Since $\chi$ is real-valued, one has $n(\overline{\psi}) = n(\psi)$. Let $m = \int_G |\chi|^2 = \sum n(\psi)^2$. One has $m \equiv \chi(1) \pmod 2$; hence $m$ is even. If $m \geqslant 4$, then it is obvious that $G$ contains an element $g$ with $|\chi(g)| \geqslant 2$. If $m < 4$, one has $m = 2$, which shows that $\chi = \varepsilon_1\psi_1 + \varepsilon_2\psi_2$ where $\psi_1$ and $\psi_2$ are distinct and $\varepsilon_1, \varepsilon_2$ belong to $\{1, -1\}$. If $\varepsilon_1 = \varepsilon_2$, take $g = 1$. If not, show that $\psi_1$ and $\psi_2$ are quadratic characters, and choose $g$ such that one of the $\psi_i$ takes the value 1 at $g$ while the other one takes the value -1.]

c) Add to the hypotheses of b) the existence of an element $\omega \in G$ such that $\chi(\omega g) = -\chi(g)$ for every $g \in G$. Show that there exist $g_+ \in G$ with $\chi(g_+) \geqslant 2$ and $g_- \in G$ with $\chi(g_-) \leqslant -2$.

[11] The restriction that $k$ is even can be deleted if $-1 \notin K$; if $-1 \in K$, one has $\mu(\varphi_k) = 0$ for every odd $k$. Note also that the constants $c$ and $c''$ are related : a simple computation shows that $c''/c = n^{d/2}\Gamma(1+d/2)$ if $-1 \notin K$ and $c''/c = 2n^{d/2}\Gamma(1+d/2)$ if $-1 \in K$.

3) Let $X_1$ and $X_2$ be two algebraic varieties over $\mathbf{Q}$. As in §7.1.3, we have the weight decompositions :

$$N_{X_1}(p) = \sum_{w \geqslant 0} (-1)^w h^w_{X_1}(p) \ \text{ and } \ N_{X_2}(p) = \sum_{w \geqslant 0} (-1)^w h^w_{X_2}(p),$$

which are valid for every $p \notin S$, for a suitable finite set $S$.

As explained in §7.1.3, each $h^w_{X_i}$ is a $\mathbf{Z}$-linear combination of $h^w_{Z_j}$, where the $Z_j$ are smooth projective varieties over $\mathbf{Q}$. In the three questions below, it is assumed that the Sato-Tate conjecture holds for all $Z_j$ and all weights $w$.

a) Let $w \in \mathbf{N}$, and define $h^w$ as $h^w_{X_1} - h^w_{X_2}$. Suppose that $h^w \neq 0$. Show that, for every $\varepsilon > 0$, the set of $p \notin S$ with $|h^w(p)| > (1-\varepsilon)p^{w/2}$ has a density which is $> 0$; in particular, this set is infinite.
[Hint. Let $K$ be the Sato-Tate group corresponding to the $Z_j$ and the weight $w$. The function $h^w(p)$ can be written as $\chi(s_p)p^{w/2}$, where $\chi$ is a character of $K$ and $s_p$ is the conjugacy class of $K$ associated with $p$ as in $(ST_3)$. Apply part a) of exerc. 2 to $K$ and $\chi$ and use the fact that the set of $x \in K$ with $|\chi(x)| > 1-\varepsilon$ is quarrable, cf. exerc. 1.]

b) Assume that $w$ is odd. Show that, for every $\varepsilon > 0$, the set of $p \notin S$ with $h^w(p) > (2-\varepsilon)p^{w/2}$ has a density which is $> 0$.
[Hint. Same method, using parts b) and c) of exerc. 2.]

c) Suppose that $|N_{X_1}(p) - N_{X_2}(p)|$ does not remain bounded when $p$ varies. Show that, for every $\varepsilon > 0$, there exist infinitely many $p$ such that

$$|N_{X_1}(p) - N_{X_2}(p)| \ > \ (2-\varepsilon)p^{\frac{1}{2}}.$$

[Hint. Use exerc.3 of §7.2.3 to show that there exists $w > 0$ such that $h^w \neq 0$; choose the largest such $w$; apply part a) of exerc. 2 if $w > 1$ and apply part b) if $w = 1$.]
Problem. Can one replace $(2-\varepsilon)p^{\frac{1}{2}}$ by $2p^{\frac{1}{2}} - O(p^{\delta})$ with $\delta < 1/2$, or maybe even with $\delta = 0$? This does not look easy, even when $X$ is an elliptic curve and $Y = \mathbf{P}_1$.

## 8.5. Examples

We give below a few explicit cases of the Sato-Tate correspondence where the weight $w$ is $0, 1$ or $2$. The case $w = 1$ is especially interesting because of the many experimental results that have been obtained recently for curves of genus 2, see §8.5.4.

### 8.5.1. The case $w = 0$ : Sato-Tate = Chebotarev

We are looking here at $H^0(X, \ell)$, where $X$ is a smooth projective variety over $\mathbf{Q}$. If $I_X$ denotes the set of the geometric connected components of $\overline{X}$, the group $\Gamma_Q$ acts on $I_X$. As in Chaper 7, let us denote by $\varepsilon_X$ the corresponding permutation character and let $G_X$ be the image of $\Gamma_Q$ in the group of permutations of $I_X$. With the notation of §8.1, we have :

$$f(p) = h^0_X(p) = \varepsilon_X(g_p) \quad \text{for } p \text{ large enough.}$$

In that case, the Sato-Tate group $K$ is the finite group $G$; the Hodge circle is trivial; the measure $\mu$ is discrete, with support equal to the set of values of $\varepsilon_X$ ; the equidistribution axiom $(A_2)$ follows from Chebotarev's Theorem 3.2, and the other axioms are obviously satisfied.

### 8.5.2. Elliptic curves with complex multiplication

We now assume $w = 1$ and we start by revisiting the case of elliptic curves, cf. §8.1.5. We begin with the CM case :

As in §8.1.5.1, $X$ is an elliptic curve over $\mathbf{Q}$ of CM-type and the function $f(p)$ is given by $f(p) = a_p/p^{\frac{1}{2}}$, with standard notation. Let us describe the Lie groups data.

8.5.2.1. *The Sato-Tate group $K$.* This group has two connected components. Its identity component $K^0$ is a circle, which can be viewed as the standard maximal torus $\mathbf{U}$ of the special unitary group[12] $\mathbf{SU}_2$; its elements can be written as diagonal matrices

$$x_\alpha = \begin{pmatrix} e^{i\alpha} & 0 \\ 0 & e^{-i\alpha} \end{pmatrix},$$

with $\alpha \in \mathbf{R}/2\pi\mathbf{Z}$. Its other connected component is made of the matrices in $\mathbf{SU}_2$ that are 0 on the diagonal :

$$\begin{pmatrix} 0 & -e^{i\alpha} \\ e^{-i\alpha} & 0 \end{pmatrix}.$$

Equivalently : *$K$ is the normalizer of $\mathbf{U}$ in $\mathbf{SU}_2$.*

The isomorphism $\mathbf{U} \to K^0$ given by

$$u \mapsto \begin{pmatrix} u & 0 \\ 0 & \overline{u} \end{pmatrix}$$

is a Hodge circle of $K$, and the elements $\omega$ and $\gamma$ of §8.2.3 are :

$$\omega = \begin{pmatrix} -1 & 0 \\ 0 & -1 \end{pmatrix} \quad \text{and} \quad \gamma = \begin{pmatrix} 0 & 1 \\ -1 & 0 \end{pmatrix}.$$

[12]This group is classically written as $\mathbf{SU}_2(\mathbf{C})$. This is a bit confusing, since $\mathbf{SU}_2$ is an algebraic group over $\mathbf{R}$ and " $\mathbf{SU}_2(\mathbf{C})$ " is the group of its real points, the group of its complex points being $\mathbf{SL}_2(\mathbf{C})$. On the other hand, writing this group as $\mathbf{SU}_2(\mathbf{R})$ could be even more confusing. This is why we use the ambiguous notation $\mathbf{SU}_2$ which does not choose between $\mathbf{R}$ and $\mathbf{C}$.

[Hence $K$ is the smallest group having a Hodge circle and an element $\gamma$ as in §8.2.3.]

8.5.2.2. *Linear representation.* It is the natural embedding $r : K \to \mathbf{SU}_2 \to \mathbf{GL}_2(\mathbf{C})$.

8.5.2.3. *Conjugacy classes.* The space $\mathrm{Cl}\, K$ has the following structure :

• the elements of $K - \mathbf{U}$ make up a unique class, with mass $\frac{1}{2}$ for the Haar measure. A representative of that class is $\gamma$.

• the conjugacy classes of $\mathbf{U}$ in $K$ can be represented by the diagonal matrices $x_\alpha$ with $0 \leqslant \alpha \leqslant \pi$ (note that $x_\alpha$ is conjugate to $x_{-\alpha}$ in $K$) ; the Haar measure on such classes is $\frac{1}{2\pi} d\alpha$ ; its mass is $\frac{1}{2}$.

8.5.2.4. *The classes $s_p$ with $p \notin S$.* These classes are defined as follows. Let $F$ be the imaginary quadratic field associated with $X$, i.e. $F = \mathbf{Q} \otimes \mathrm{End}_{\overline{\mathbf{Q}}}(X)$, and let $\varepsilon_F : \Gamma_S \to \{\pm 1\}$ be the corresponding quadratic character. Then :

• if $p$ is inert in $F/\mathbf{Q}$, i.e. $\varepsilon_F(p) = -1$, then $s_p$ is equal to the class $\gamma$ above ;

• if $p$ splits in $F/\mathbf{Q}$, i.e. $\varepsilon_F(p) = 1$, then $s_p$ is the class of the elements $x_\alpha$ such that $2 \cos \alpha = f(p)$.

8.5.2.5. *The axioms.* Axiom $(A_0)$ is obviously satisfied : the map $K \to \mathbf{GL}_2(\mathbf{C})$ is an embedding.

To check $(A_1)$, we have to show that $\mathrm{Tr}(s_p) = f(p)$ ; this is clear by construction if $p$ splits in $F$ ; if $p$ does not, then it is well known that $f(p) = 0$, hence $f(p) = \mathrm{Tr}(\gamma)$ as required.

We have already given in 8.1.5.1 the references to Hecke and Deuring for the non-trivial, but well-known, equidistribution axiom $(A_2)$.

8.5.2.6. *The group $K/K^0$ as a Galois group.* Since this group has two elements, one may identify it with $\{\pm 1\}$, and we thus get, as in §8.5.2.4, a surjective homomorphism $\varepsilon_F : \Gamma_S \to K/K^0$. Axiom $(A_3)$ is obviously satisfied.

8.5.2.7. *Other axioms.* Axiom $(A_4)$ holds since the quadratic field $F$ is imaginary. As for $(A_5)$, it follows from the fact that $K$ is contained in $\mathbf{SL}_2(\mathbf{C})$.

*Exercise.* Use 8.2.3.6 ii) to recover the well-known fact that $X$ has potential good reduction everywhere, i.e. that its $j$-invariant is an integer.

### 8.5.3. Elliptic curves without complex multiplication

The notation and the hypotheses are the same as in §8.5.2, except that the endomorphism ring of $X_{/\overline{\mathbf{Q}}}$ is now supposed to be $\mathbf{Z}$.

The Sato-Tate group $K$ is $\mathbf{SU}_2$ : it can't be anything else since the only subgroups of $\mathbf{SU}_2$ containing the maximal torus $\mathbf{U}$ are $\mathbf{U}$ itself, its normalizer $N(\mathbf{U}) = \langle \mathbf{U}, \gamma \rangle$, and $\mathbf{SU}_2$; moreover, the first two possibilities are eliminated because (for instance) they are not compatible with the size of the $\ell$-adic representations.

The Hodge circle and the elements $\omega$ and $\gamma$ are the same as in the CM case.

The space $\mathrm{Cl}\, K$ may be identified with $[0, \pi]$ in such a way that the conjugacy class of $x_\alpha \in T$ corresponds to $\alpha$ if $0 \leqslant \alpha \leqslant \pi$. The Haar measure is known to be $\frac{2}{\pi} \sin^2\alpha \, d\alpha$. The trace map $\mathrm{Tr} : K \to I = [-2, 2]$ gives a homeomorphism of $\mathrm{Cl}\, K$ onto $I$, which maps the class of $x_\alpha$ to $z = 2 \cos x_\alpha$.

If $p \notin S$, the element $s_p$ of $\mathrm{Cl}\, K$ is defined by the condition that its trace is $f(p) = a_p/p^{\frac{1}{2}}$.

The axioms of §8.2 are satisfied. This is clear for all of them, except for $(A_2)$ (the original "Sato-Tate conjecture"), for which we refer to the bibliographical indications given in §8.1.5.2; note that some of them apply not only over $\mathbf{Q}$ but also over any totaly real number field.

The equidistribution measure is the one given in §8.1.5.2, namely

$$\mu = \frac{1}{2\pi}\sqrt{4 - z^2}\, dz \quad = \quad \frac{2}{\pi} \sin^2\alpha \, d\alpha.$$

For a picture of its density, computed using the values of $f(p)$ in a random numerical case[13], see [KS 09, §9.1, $s_1$-diagram]; note that the other diagrams of [KS 09, §9.1] represent the densities of the measures $\mu_e$ $(e = 2, 3, 4)$ introduced in the exercise at the end of §8.1.5.

### 8.5.4. Two elliptic curves

Instead of one elliptic curve, we may take two (or more) of them. Consider for instance the case of two elliptic curves $X_1$ and $X_2$. Let $K_1$ and $K_2$ be the corresponding Sato-Tate groups. The Sato-Tate group $K$ of the pair $(X_1, X_2)$ is a closed subgroup of $K_1 \times K_2$; moreover both projections $K \to K_1$ and $K \to K_2$ are surjective. There are several cases :

a) None of the curves are of CM-type, so that $K_1$ and $K_2$ are both equal to $\mathbf{SU}_2$. This case subdivides in three :

• $K = K_1 \times K_2$; this occurs when $X_1$ and $X_2$ are not $\overline{\mathbf{Q}}$-isogenous. We have $K = \mathbf{SU}_2 \times \mathbf{SU}_2$.

• $K$ is the graph of an isomorphism of $K_1$ onto $K_2$; this occurs where $X_1$ and $X_2$ are $\mathbf{Q}$-isogenous. We have $K = \mathbf{SU}_2$, embedded diagonally in $\mathbf{SU}_2 \times \mathbf{SU}_2$.

[13] The equation of $X$ chosen in [KS 09] is $y^2 = x^3 + 314159x + 271828$, which is random enough for our problem, even though its coefficients have a familiar look.

• $K = C_2 \cdot \mathbf{SU}_2$, where $\mathbf{SU}_2$ is embedded diagonally, and $C_2$ is the center of $\{1\} \times \mathbf{SU}_2$; this occurs when $X_2$ is isogenous to a non-trivial quadratic twist of $X_1$.

b) The first curve is of CM-type and the second one is not. We then have $K = K_1 \times K_2$, i.e. $K = \mathbf{SU}_2 \times N(\mathbf{U})$.

c) Both curves are of CM-type, and their fields of complex multiplication are different. We have $K = K_1 \times K_2 = N(\mathbf{U}) \times N(\mathbf{U})$. The group $K/K^0$ is elementary of type (2,2).

d) Both curves are of CM-type with the same field of complex multiplication. We then have $K = C_n \cdot N(\mathbf{U})$, where $N(\mathbf{U})$ is embedded diagonally, $n$ is an integer equal to 1, 2, 3, 4 or 6, and $C_n$ is a cyclic subgroup of $\mathbf{U} \times \mathbf{U}$, of order $n$ and intersecting trivially $K^0$. The group $K/K^0$ is dihedral of order $2n$. One has $n = 1$ if and only if the two curves are $\mathbf{Q}$-isogenous. As for $n = 2, 3, 4, 6$, see exerc.1 and 2.

In each of the cases a), b), c), d), the Lie groups data are defined in an obvious way. Axiom $(A_2)$ is satisfied in cases c) and d), where there is potential complex multiplication. I am not sure whether it has already been proved in cases a) and b). In the exercises below, we assume that it has been.

*Exercises.*

1) Show that case d) with $n = 4$ may only occur when the field of complex multiplication is $\mathbf{Q}(i)$. If that field is $\mathbf{Q}(i)$, show that $X_1$ and $X_2$ are $\mathbf{Q}$-isogenous to curves defined by the equations $y^2 = x^3 + a_1 x$ and $y^2 = x^3 + a_2 x$ with $a_1, a_2 \in \mathbf{Q}^\times$. Let $\lambda = a_1/a_2$ and let $F = \mathbf{Q}(i, \lambda^{\frac{1}{4}})$. Show that $n = \frac{1}{2}[F : \mathbf{Q}]$ and $K/K^0 \simeq \mathrm{Gal}(F/\mathbf{Q})$. In particular, one has $n = 1$ if and only if either $\lambda$ or $-4\lambda$ is a 4-th power in $\mathbf{Q}$.

2) Show that case d) with $n = 3$ or 6 may only occur when the field of complex multiplication is $\mathbf{Q}(\sqrt{-3})$. If that field is $\mathbf{Q}(\sqrt{-3})$, show that $X_1$ and $X_2$ are $\mathbf{Q}$-isogenous to curves defined by the equations $y^2 = x^3 + b_1$ and $y^2 = x^3 + b_2$ with $b_1, b_2 \in \mathbf{Q}^\times$. Let $\lambda = b_1/b_2$ and let $F = \mathbf{Q}(\sqrt{-3}, \lambda^{\frac{1}{6}})$. Show that $n = \frac{1}{2}[F : \mathbf{Q}]$ and $K/K^0 \simeq \mathrm{Gal}(F/\mathbf{Q})$. In particular, one has $n = 1$ if and only if either $\lambda$ or $-27\lambda$ is a 6-th power in $\mathbf{Q}$.

3) Assume that $X_1$ and $X_2$ are not $\mathbf{Q}$-isogenous, and exclude case d) with $n = 3$. Show that the central element $(1, -1)$ of $\mathbf{SU}_2 \times \mathbf{SU}_2$ belongs to $K$. Deduce from this, and from $(A_2)$, that, for every $\varepsilon > 0$, there are infinitely many $p$ such that $N_{X_1}(p) - N_{X_2}(p) \geqslant (4 - \varepsilon)p^{\frac{1}{2}}$; more generally, the ratios $(N_{X_1}(p) - N_{X_2}(p))/p^{\frac{1}{2}}$ are equidistributed with respect to a measure whose support is equal to $[-4, 4]$.

In case d) with $n = 3$, prove a similar result with 4 replaced by $2\sqrt{3}$.

[Note that the bound so obtained is better than the one given in exerc. 3 of §8.4.4.]

### 8.5.5. Curves[14] of genus 2

Here, there are few proofs (for the time being), but there is a wealth of numerical examples and computations, thanks to Kedlaya-Sutherland [KS 09], completed in [FKRS 11].

The main difference with elliptic curves is that there are many more possibilities for the Sato-Tate group $K$ ; instead of merely 2, there are 34 of them, whose detailed description can be found in [FKRS 11] ; this number would even be larger if the ground field, instead of being $\mathbf{Q}$, were an arbitrary number field.

The generic case, similar to $\mathbf{SU}_2$ for genus 1, is $K = \mathbf{USp}_4$ , a maximal compact subgroup of the symplectic group $\mathbf{Sp}_4(\mathbf{C})$ ; this case occurs if and only if the Jacobian variety $J$ of the given curve $X$ "has no endomorphism", i.e. $\mathrm{End}_{/\overline{\mathbf{Q}}}(J) = \mathbf{Z}$, see below. It gives a measure $\mu$ on the segment $[-4, 4]$ whose density is maximum at $z = 0$ and tends to 0 very fast when $|z| \to 4$ ; indeed, that density is asymptotically $\frac{1}{64\pi}\varepsilon^4$ when $|z| = 4 - \varepsilon$, with $\varepsilon \to 0$ and $\varepsilon > 0$. The moments $\mu(\varphi_k)$ of $\mu$ are computed in [KS 03, §4.1] ; the first ones are [1,0,1,0,3,0,14,0,84,0,594,0,4719]. They are 0 when $k$ is odd, and when $k$ is even they are given by the formula :

$$\mu(\varphi_{2m}) = C_m C_{m+2} - C_{m+1}^2 \quad \text{for } m \geqslant 0,$$

where $C_m$ is the $m$-th Catalan number. A more explicit formula, due to A. Mihailovs (quoted in [OEIS A005700]), is :

$$\mu(\varphi_{2m}) = \frac{6.\, 2m!\, (2m+2)!}{m!\, (m+1)!\, (m+2)!\, (m+3)!} \quad \text{for } m \geqslant 0,$$

which implies

$$\mu(\varphi_{2m}) \sim \frac{24}{\pi} 4^{2m}/m^5 \quad \text{when } m \to \infty.$$

(Note the exponent 5 in $m^5$, which is equal to $\frac{1}{2}\dim K$, cf. §8.4.4.4.)

The Sato-Tate group $K$ associated with a given curve $X$ of genus 2 is a closed subgroup of $\mathbf{USp}_4$. Moreover, by 8.2.3.6 i), its identity component $K^0$ is generated by Hodge circles. Using the list of the connected reductive subgroups of $\mathbf{Sp}_4$, one finds that there are six possibilities for $K^0$. They correspond to the different possible structures of the $\mathbf{Q}$-algebra $A = \mathbf{Q} \otimes \mathrm{End}_{/\overline{\mathbf{Q}}} J$, where $J$ is the Jacobian of $X$, cf. e.g. [Mo 04, §5.7]. If one orders them by decreasing dimension of $K$, they are :

[14] Everything we say below applies more generally to abelian varieties of dimension 2. Note that there are examples of such varieties that are not $\mathbf{Q}$-isogenous to any principally polarized one, cf. [Ho 01] ; their Sato-Tate group cannot thus be reduced to that of a curve of genus 2.

i) $K^0 = \mathbf{USp}_4$ ; $\dim K = 10$ ; $A = \mathbf{Q}$ ;

ii) $K^0 = \mathbf{SU}_2 \times \mathbf{SU}_2$ ; $\dim K = 6$; $A$ is commutative and étale of rank 2 : either $\mathbf{Q} \times \mathbf{Q}$ or a real quadratic field ;

iii) $K^0 = \mathbf{U} \times \mathbf{SU}_2$ ; $\dim K = 4$; $A = \mathbf{Q} \times E$, where $E$ is an imaginary quadratic field ;

iv) $K^0 = \mathbf{SU}_2$, diagonally embedded in $\mathbf{SU}_2 \times \mathbf{SU}_2$ ; $\dim K = 3$; $A$ is a quaternion algebra over $\mathbf{Q}$ that splits over $\mathbf{R}$ (it may also split over $\mathbf{Q}$) ;

v ) $K^0 = \mathbf{U} \times \mathbf{U}$ = maximal torus of $\mathbf{USp}_4$ ; $\dim K = 2$ ; $A$ is commutative and étale of rank 4 (if $A$ is a field, it is a primitive CM-field and the Galois group of its Galois closure is either cyclic of order 4, or dihedral of order 8, cf. [Sh 98, p. 64] ; if $A$ is not a field, it is a product of two non-isomorphic imaginary quadratic fields) ;

vi) $K^0 = \mathbf{U}$, diagonally embedded in $\mathbf{U} \times \mathbf{U}$ ; $\dim K = 1$ ; $A$ is isomorphic to the matrix algebra $\mathbf{M}_2(E)$ where $E$ is an imaginary quadratic field.

Examples of each case are easily constructed :

For the first one, one may take the curve defined by the equation $y^2 = x^5 + x + 1$, cf. [KS 09, Table 11] ; since the Galois group over $\mathbf{Q}$ of the polynomial $x^5 + x + 1$ is the symmetric group $S_5$, it follows from a general theorem of Zarhin (cf. [Za 02]) that $A = \mathbf{Q}$, hence the Sato-Tate group is $\mathbf{USp}_4$.

For the other five, one chooses a curve whose jacobian is isogenous to a product of two elliptic curves of suitable types ; see [KS 09] and [FKRS 11] for explicit examples.

What is more difficult is to find the list of possibilities for $K$ when $K^0$ is given, because the component group $K/K^0$ can be rather large, especially in case vi), where it can have as many as 48 elements, see below.

*The* $\mathbf{SO}_5$ *dictionary.*

The group $K$ contains the central element $-1$ of $\mathbf{USp}_4$ ; hence it is well defined by its image $K'$ in the quotient $\mathbf{USp}_4/\{\pm 1\}$ which is isomorphic[15] to $\mathbf{SO}_5(\mathbf{R})$. A conjugacy class $z$ of $K$ thus gives a conjugacy class $g$ in $\mathbf{SO}_5(\mathbf{R})$, i.e. two angles $(u, v)$, defined up to sign and up to permutation. If $a_1$ and $a_2$ are the first two coefficients of the characteristic polynomial of $g$, we have $a_1^2 = 4(1 + \cos u)(1 + \cos v)$ and $a_2 = 2(1 + \cos u + \cos v)$. This dictionary makes the description of $K$ easier. For instance, case vi) corresponds to $K'^0 = \mathbf{SO}_2(\mathbf{R})$, embedded in $\mathbf{SO}_5(\mathbf{R})$ in an obvious way.

[15] We are taking advantage of the fact that the Lie types $B_g$ and $C_g$ coincide when $g = 2$. No such luck when $g > 2$ !

By looking at the normalizer of $\mathbf{SO}_2(\mathbf{R})$ in $\mathbf{SO}_5(\mathbf{R})$, one sees that the finite group $F = K/K^0 = K'/K'^0$ embeds in $\{\pm 1\} \times \mathbf{SO}_3(\mathbf{R})$ ; moreover, the elements of $F$ have order 1, 2, 3, 4 or 6 : this follows from the rationality properties of the Frobenius elements $s_p$, combined with the equations for $a_1^2$ and $a_2$ given above. These properties imply that $F$ is a subgroup of either $\{\pm 1\} \times D_6$ or $\{\pm 1\} \times S_4$, where $D_6$ is the dihedral group of order 12, and $S_4$ is the symmetric group on 4 letters ; moreover, the projection $F \to \{\pm 1\}$ is surjective, because of the element $\gamma$ of §8.2.3.4 (complex conjugation). As shown in [FKRS 11], all the finite groups satisfying these conditions do occur, with three exceptions which would not be compatible with the known structure of the algebra $A$.

### 8.5.6. The case $w = 2$ : projective surfaces

Here, $X$ is a smooth projective surface over $\mathbf{Q}$ and $w = 2$, i.e. we look at $H^2(X, \ell)$ ; we assume for simplicity that $X$ is geometrically irreducible, and we put $n = B_2(X) = h^{2,0} + h^{1,1} + h^{0,2}$, where the $h^{p,q}$ are the Hodge numbers :

$$h^{2,0} = \dim H^2(X, \Omega^0) = \dim H^0(X, \Omega^2) = h^{0,2} \text{ and } h^{1,1} = \dim H^1(X, \Omega^1),$$

with standard notation. The Hodge circle has weights $\{-2, 0, 2\}$ with multiplicity $h^{0,2}, h^{1,1}, h^{2,0}$ respectively.

If $\rho$ is the Picard number of $\overline{X}$, i.e. the rank of the Néron-Severi group $NS$, we have $\rho \leqslant h^{1,1}$.

8.5.6.1. *The embedding $K \subset \mathbf{O}_n(\mathbf{R})$ and the character $\varepsilon$.*

Since $w$ is even, the Sato-Tate group $K$ leaves invariant a non-degenerate quadratic form, cf. §8.2.3.5 ; this gives an embedding of $K$ into the orthogonal group $\mathbf{O}_n(\mathbf{R})$. This embedding depends on the choice of a polarization on $X$, but the corresponding quadratic character

$$\varepsilon : K \to \mathbf{O}_n(\mathbf{R}) \to \{1, -1\}$$

is independent of it ; since it is trivial on $K^0$, we may view it as a character of $\Gamma_{\mathbf{Q}}$ : it corresponds to the action of that group on $\wedge^n H^2(\overline{X}, \mathbf{Q}_\ell(1))$. (Note the Galois twist by $w/2 = 1$ ; it has the same effect as the analytic normalisation $f(p) = h^w(p)/p^{w/2}$ of §8.1.1.)

The character $\varepsilon$ can be described as a kind of discriminant for the de Rham cohomology of $X$, cf. [Sa 94, Theorem 2].

8.5.6.2. *The embedding $K^0 \subset \mathbf{O}_{n-\rho}(\mathbf{R})$.*

We have a natural embedding of $NS \otimes_{\mathbf{Z}} \mathbf{Q}_\ell$ into $H^2(\overline{X}, \mathbf{Q}_\ell(1))$, and the action of $\Gamma_{\mathbf{Q}}$ on that subspace factors through a finite group. This implies that the identity component $K^0$ of $K$ acts trivially on a $\rho$-dimensional space, i.e. is contained in a conjugate of $\mathbf{O}_{n-\rho}(\mathbf{R})$.

Note also that $NS \otimes_{\mathbf{Z}} \mathbf{Q}_\ell$ contains a 1-dimensional subspace fixed under $\Gamma_{\mathbf{Q}}$ ; this shows that $K$ is contained in $\mathbf{O}_{n-1}(\mathbf{R})$, up to conjugation.

8.5.6.3. *The case where* $h^{2,0} = 0$.

Here the Hodge circle is trivial, and so is $K^0$ ; indeed, by a well-known theorem of Lefschetz, we have $\rho = h^{1,1} = n$ : the whole cohomology is algebraic. Hence $K$ is finite.

This applies in particular when $X$ is geometrically rational, cf. §2.3.3. For instance, if $X$ is a smooth cubic surface, $K$ is the image of $\Gamma_{\mathbf{Q}} \to \mathrm{Weyl}(E_6)$, and the quadratic character $\varepsilon$ is $\Gamma_{\mathbf{Q}} \to \mathrm{Weyl}(E_6) \to \{1,-1\}$ ; this character can be written explicitly in terms of the discriminant of an equation of $X$ [16] , cf. [EJ 10, Theorem 2.12].

8.5.6.4. *The case where X is an abelian surface.*

We have $n = 6$, so that $K$ may be viewed (see §8.5.6.2) as a subgroup of $\mathbf{O}_5(\mathbf{R})$. In fact, $K$ is a subgroup of $\mathbf{SO}_5(\mathbf{R})$. More precisely, it is the group denoted by $K'$ at the end of §8.5.5 ; according to [FKRS 11], there are 34 different possibilities for it.

8.5.6.5. *$K3$-surfaces.*

If $X$ is a $K3$-surface, we have $h^{2,0} = h^{0,2} = 1, h^{1,1} = 20$, and $n = 22$. The Picard number $\rho$ belongs to the interval [1,20]. The two extreme cases $\rho = 1$ and $\rho = 20$ are possible :

i) The generic case is $\rho = 1$. In that case, it may happen that $K$ is equal to $\mathbf{O}_{21}(\mathbf{R})$, cf. [Te 85] and [El 04].

ii) The case $\rho = 20$ implies $K^0 = \mathbf{SO}_2(\mathbf{R}) =$ Hodge circle ; there is potential complex multiplication. Explicit examples can be found in [SI 77], [PTV 92] and [Li 95].

*Exercise.* Let $\varepsilon$ be the quadratic character of $\Gamma_{\mathbf{Q}}$ defined in §8.5.6.1 ; view it as a Dirichlet character with values in $\{1,-1\}$.

a) Show that, for every large enough $p$, the sign of the functional equation of $\zeta_{X,p}$ is $(-1)^n\varepsilon(p)$, where $n = B_2(X)$.

b) Show that $\varepsilon(-1) = (-1)^m$, where $m = \frac{1}{2}(N_X(1)+N_X(-1)) = \frac{1}{2}(\chi(X(\mathbf{C})+\chi(X(\mathbf{R})))$.

[16] More precisely, one has $\varepsilon(p) = (\frac{-3\Delta}{p})$, with the definition of the discriminant $\Delta$ given by G. Salmon in 1861 and corrected by W.L. Edge in 1980, as explained in [EJ 10, §2]. There is a similar formula for every even-dimensional smooth hypersurface of a projective space, cf. [SS 12].

CHAPTER 9

# HIGHER DIMENSION : THE PRIME NUMBER THEOREM AND THE CHEBOTAREV DENSITY THEOREM

As we shall see, the results of §7.2.1 can be reformulated as a kind of "prime number theorem" for a scheme $T$ of arbitrary dimension; one may then define a notion of density (cf. [Pi 97 App.B]) and prove a "Chebotarev density theorem" as in [Se 65], [Fa 84] and [Pi 97]. One can then transpose to the case of a $T$-scheme $X$ almost everything done in the previous chapters for a $\mathbf{Z}$-scheme.

## 9.1. The prime number theorem

### 9.1.1. Notation

Let $T$ be a scheme of finite type over $\mathbf{Z}$; as in §1.5, we denote by $\underline{T}$ the set of its closed points. If $t \in \underline{T}$, the residue field $\kappa(t)$ is finite; let $|t|$ be its number of elements and $p_t$ its characteristic. We have $|t| = p_t^{d(t)}$ with $d(t) \geqslant 1$; the integer $d(t)$ is called the *degree* of $t$.

For any given $x \in \mathbf{R}$ there are only finitely many $t$ with $|t| \leqslant x$; let $\pi_T(x)$ be their number. When $T = \operatorname{Spec} O_K$, where $O_K$ is the ring of integers of a number field $K$, the function $\pi_T$ coincides with the prime counting function $\pi_K$ of §3.1.1.

### 9.1.2. Hypotheses

We shall be interested in the case where $T$ has the following two properties :

($a$) $T$ is irreducible; we denote by $d$ its dimension, and by $K$ the residue field of its generic point.

($b$) The natural map $T \to \operatorname{Spec} \mathbf{Z}$ is dominant (equivalently : the field $K$ has characteristic 0, in which case it is an extension of $\mathbf{Q}$ of transcendence degree $d-1$).

Note that we are eliminating the equal characteristic case, where $K$ has characteristic $p > 0$. This case is interesting, too, but the statements are a bit different : the function $\pi_T(x)$ jumps so much when $x$ reaches a power of $p$ that none of the theorems 9.1 and 9.11 below remains valid - they have to be restated in terms of Dirichlet's analytic density, as in [Se 65] and [Pi 97].

### 9.1.3. Statement of the theorem

We make on $T$ the two hypotheses (a) and (b) above. The following statement is a generalization of Theorem 3.1 :

**Theorem 9.1.** *There exists $c > 0$ such that*

$$\pi_T(x) = \mathrm{Li}(x^d) + O(x^d \exp(-c\sqrt{\log x}\,)) \quad \textit{for} \quad x \to \infty.$$

**Corollary 9.2.** $\pi_T(x) \sim \frac{x^d}{d \log x}$ *for* $x \to \infty$.

The proof of Theorem 9.1 will be given in §9.1.6.

*Remark.* Under GRH, the error term in Theorem 9.1 can be improved to $O(x^{d-\frac{1}{2}} \log x)$ ; the proof is the same.

### 9.1.4. Reduction to degree 1

We now show that the $t$ of degree $> 1$ can be neglected. More precisely, let us define :

$\pi_T^1(x)$ = number of $t \in \underline{T}$ of degree 1 with $|t| \leqslant x$ ;

$\pi_T^2(x)$ = number of $t \in \underline{T}$ of degree $\geqslant 2$ with $|t| \leqslant x$.

We have $\pi_T(x) = \pi_T^1(x) + \pi_T^2(x)$. The following lemma shows that $\pi_T^2(x)$ is indeed negligible, compared with $\mathrm{Li}(x^d)$ :

**Lemma 9.3.** $\pi_T^2(x) = O(x^{d-\frac{1}{2}})$ *for* $x \to \infty$.

*Remark.* This bound is not optimal; the order of magnitude of $\pi_T^2(x)$ is in fact $x^{d-\frac{1}{2}}/\log x$ : see the Exercise at the end of the next section.

### 9.1.5. Proof of Lemma 9.3

Note first that every $t \in \underline{T}$ of degree $e$ gives $e$ elements of $T(\mathbf{F}_{p^e})$, with $p = p_t$ ; in particular, we have

$$\pi_T^2(x) \leqslant \sum_{p,e} N_T(p^e)/e,$$

where the sum is over the pairs $(p, e)$ with $2 \leqslant e \leqslant \frac{\log x}{\log p}$. The $p$-fiber of $T$ is an $\mathbf{F}_p$-scheme of dimension $\leqslant d-1$, and its Betti numbers with proper support remain bounded when $p$ varies (this is a general property of algebraic families - it follows from the constructibility of the direct image sheaves $R^i f_! \mathbf{Q}_\ell$, where $f$ is the projection $T \to \mathrm{Spec}\, \mathbf{Z}$). This implies (cf. Theorem 4.7) :

$$N_T(p^e) \ll p^{e(d-1)} \ll x^{d-1}.$$

The number of $e$'s is bounded by $\log x/\log p \ll \log x$ and the number of $p$'s is $\pi(x^{\frac{1}{2}}) \ll x^{\frac{1}{2}}/\log x$. We thus get :

$$\pi_T^2(x) \ll x^{d-1}x^{\frac{1}{2}} \log x/\log x = x^{d-\frac{1}{2}},$$

as wanted.

*Exercise.* Let $\varepsilon_T$ be the permutation character of $\Gamma_{\mathbf{Q}}$ associated with $T$ as in §7.2.1, and let $\Psi^2\varepsilon_T$ be its transform by the Adams operator $\Psi^2$. Show that

$$\pi_T^2(x) \sim \frac{<\Psi^2\varepsilon_T, 1> x^{d-\frac{1}{2}}}{(2d-1)\log x} \quad \text{for } x \to \infty.$$

[Hint. Use Corollary 7.11 to estimate the terms corresponding to $e = 2$, and show that the contribution of the others is $\ll x^{d-\frac{2}{3}}$.]

### 9.1.6. Proof of Theorem 9.1

Because of Lemma 9.3, we may neglect the $t$'s of degree $> 1$, i.e. replace $\pi_T(x)$ by $\pi_T^1(x)$. But $\pi_T^1(x)$ is the same as the sum $\sum_{p \leqslant x} N_T(p)$. By Corollary 7.1.3, applied to $T$ with $d$ replaced by $d-1$, there exists $c > 0$ such that

$$\pi_T^1(x) = \mathrm{Li}(x^d) + O(x^d e^{-c\sqrt{\log x}}) \quad \text{for} \quad x \to \infty.$$

### 9.1.7. The zeta point of view

Instead of counting points, one may look at the zeta function $\zeta_T(s) = \prod_{t \in \underline{T}} 1/(1 - |t|^{-s})$, as in §1.5; the product converges absolutely for $\mathrm{Re}(s) > d$.

It turns out that this function does not differ much from $\zeta_{K_0}(s-d+1)$, where $K_0$ is the maximal number field contained in $K$ (i.e. the algebraic closure of $\mathbf{Q}$ in $K$). In order to state the result more precisely, let us say that an Euler product

$$F = \prod_p \prod_i (1 - \alpha_{i,p} p^{-s})^{-1}$$

has *size* $\leqslant r$, where $r$ is a positive real number, if :

i) the number of the indices $i$ relative to a given $p$ remains bounded when $p$ varies ;

ii) $|\alpha_{i,p}| \leqslant p^r$ for every pair $(i, p)$.

Such an Euler product converges absolutely for $\mathrm{Re}(s) > r+1$. The relation between $\zeta_T(s)$ and $\zeta_{K_0}(s-d+1)$ can then be stated as :

**Theorem 9.4.** *There exists two Euler products $F_1$ and $F_2$ of size $\leqslant d - \frac{3}{2}$ such that $\zeta_T(s) = \zeta_{K_0}(s-d+1)F_1(s)^{-1}F_2(s).m(s)$, where $m(s)$ is a finite product of factors $(1 - \lambda p^{-s})^{\pm 1}$ with $|\lambda| = p^{d-1}$.*

The proof is analogous to that of Theorem 9.1; it is given, in a slightly less precise form, in the three papers [Se 65], [Fa 84] and [Pi 86] quoted above. One may summarize it by saying that the top cohomology (in dimension $2d-2$) of the fibers of $T \to \operatorname{Spec} \mathbf{Z}$ gives the term $\zeta_{K_0}(s-d+1)m(s)$, while the lower cohomology gives $F_1$ and $F_2$. The proof also shows that one can choose $F_1, F_2$ and $m$ such that all the $\alpha_{i,p}$ that occur in $F_1$ and $F_2$ are $p$-Weil integers of weight $\leqslant 2d-3$ and that the $\lambda$'s occurring in $m(s)$ are Weil numbers of weight $2d-2$.

**Corollary 9.5** (cf [Se 65]). *The function $\zeta_T(s)$ extends analytically to a meromorphic function on the half-plane* $\mathrm{Re}(s) \geqslant d - \frac{1}{2}$ *that has a simple pole at $s = d$ and is holomorphic elsewhere.*

Moreover any zero-free region for $\zeta_{K_0}(s)$ inside $\mathrm{Re}(s) > \frac{1}{2}$ gives a corresponding zero-free region for $\zeta_T(s)$ inside $\mathrm{Re}(s) > d - \frac{1}{2}$. In particular, $\zeta_T(s)$ is holomorphic and non-zero at every point $s \neq d$ of the line $\mathrm{Re}(s) = d$.

*Remark.* One expects also that the functions $F_1$ and $F_2$ are finite products of functions of the form $f_i(s - \frac{1}{2}c_i)$, where $f_i(s)$ belongs to the Selberg class (cf. e.g. [KP 99]), and $c_i$ is an integer with $0 \leqslant c_i \leqslant 2d-3$.

## 9.2. Densities

We keep the hypotheses of §9.1.2 : the scheme $T$ is irreducible of dimension $d$ and the map $T \to \operatorname{Spec} \mathbf{Z}$ is dominant.

### 9.2.1. Definition

Let $P$ be subset of $\underline{T}$; as in §3.1.3, if $x$ is a real number, we denote by $\pi_P(x)$ the number of $t \in \underline{T}$ with $|t| \leqslant x$, and we define the *upper density* and the *lower density* of $P$ by the formulae :

$$\text{upper-dens}(P) = \limsup \pi_P(x)/\pi_T(x) \quad \text{for } x \to \infty,$$

and

$$\text{lower-dens}(P) = \liminf \pi_P(x)/\pi_T(x) \quad \text{for } x \to \infty.$$

We say that $P$ *has density* $\lambda$ if upper-dens$(P)$ = lower-dens$(P) = \lambda$, i.e. if

$$\pi_P(x) = \lambda x^d / d \log x + o(x^d/\log x) \quad \text{for} \quad x \to \infty.$$

*Numbering.* It is sometimes convenient to restate the above definitions by numbering the elements of $\underline{T}$ as $\{t_1, ..., t_n, ...\}$ with the following two conditions :

(i) $t_i = t_j \iff i = j$;

(ii) $|t_i| < |t_j| \implies i < j$.

A subset $P$ of $\underline{T}$ has density $\lambda$ in $\underline{T}$ if and only if the set of $n$'s with $t_n \in \underline{T}$ has density $\lambda$ in $\mathbf{N}$.

[That this notion is independent of the numbering follows from the fact that two different numberings $\{t_1, ..., t_n, ...\}$ and $\{t'_1, ..., t'_n, ...\}$ are close to each other in the following sense : if $\sigma$ is the unique permutation of $\mathbf{N}$ such that $t'_n = t_{\sigma(n)}$, then $|n - \sigma(n)| = o(n)$. Such a result would not hold in the equicharacteristic case.]

### 9.2.2. Examples of sets of density 0

**Proposition 9.6.** *If $P$ is not Zariski-dense in $T$, then* $\pi_P(x) = O(x^{d-1}/\log x)$; *in particular* $\mathrm{dens}(P) = 0$.

**Corollary 9.7.** *If* upper-dens$(P) > 0$, *then $P$ is Zariski-dense in $T$.*

*Proof of Proposition 9.6.* Since $P$ is not Zariski-dense, it is contained in a finite union of $\underline{T}_i$, with $T_i \subset T$ and $\dim T_i < d$. By Corollary 9.2, each $\pi_{T_i}(x)$ is $O(x^{d-1}/\log x)$. This implies $\pi_P(x) = O(x^{d-1}/\log x)$; hence $\mathrm{dens}(P) = 0$.

**Proposition 9.8.** *The subset of $\underline{T}$ made up of the $t$'s of degree $> 1$ has density* 0.

*Proof.* This follows from Lemma 9.3.

### 9.2.3. Birational invariance of the density

Let $T'$ be another irreducible scheme of finite type over $\mathrm{Spec}\,\mathbf{Z}$, and assume that it is "birationally isomorphic" to $T$, i.e. that we have open dense subschemes $U$ and $U'$ of $T$ and $T'$ and an isomorphism $U \simeq U'$ by which we identify $U$ and $U'$.

**Proposition 9.9.** *Let $P$ be a subset of $\underline{T}$ and let $P'$ be a subset of $\underline{T}'$. Suppose that $P' \cap \underline{U}' = P \cap \underline{U}$. Then* $\mathrm{dens}(P) = \mathrm{dens}(P')$.

*Proof.* Let $Q = P - P \cap \underline{U}$ and let $Q' = P' - P' \cap \underline{U}'$. Proposition 9.6 shows that $\pi_Q(x) = O(x^{d-1}/\log x)$, hence $\mathrm{dens}(Q) = 0$; similarly $\pi'_Q(x) = O(x^{d-1}/\log x)$, and $\mathrm{dens}(Q') = 0$. Hence :

$$\mathrm{dens}(P) = \mathrm{dens}(P \cap \underline{U}) = \mathrm{dens}(P' \cap \underline{U}') = \mathrm{dens}(P').$$

The three propositions above will allow us to make several restrictive assumptions on $T$, if needed. For instance, we shall be able to assume that $T$ is reduced, or normal, or smooth over $\mathbf{Z}$, and that it is affine. We shall also be able to restrict ourselves to closed points of degree 1.

*Exercise.* If $P$ and $P'$ are as in Proposition 9.9, show that $\pi_P(x) - \pi_{P'}(x) = O(x^{d-1}/\log x)$.

### 9.2.4. Example of a set of density $\frac{1}{2}$

Suppose we have a dominant map $f : T' \to T$, where $T'$ has the same properties as above, with the difference that $f$ is "quadratic", i.e. gives rise to a quadratic extension $K'/K$, where $K'$ is the residue field of the generic point of $T'$.

**Proposition 9.10.** *Let $P$ be the subset of $\underline{T}$ made up of the points $t$ such that there exists $t \in \underline{T}'$ such that $f(t') = t$ and $d(t') = d(t)$. Then* $\operatorname{dens}(P) = \frac{1}{2}$.

[This may be viewed as a special case of the Chebotarev theorem discussed in the next section.]

*Proof.* Using the reductions mentioned in §9.2.3, we may assume that the map $f$ is finite étale of degree 2. Let $\underline{T}^1$ and $\underline{T}'^1$ be the subsets of $\underline{T}$ and $\underline{T}'$ made up of the points of degree 1. The image of $f : \underline{T}'^1 \to \underline{T}^1$ is $P \cap \underline{T}^1$, and each fiber of that map has two elements. This shows that $\pi^1_{\underline{T}'}(x) = 2\pi_{P\cap\underline{T}^1}(x)$ for every $x \in \mathbf{R}$. By Theorem 9.1 and Lemma 9.3, we have $\pi^1_{\underline{T}'}(x) \sim \operatorname{Li}(x^d)$, hence $\pi_{P\cap\underline{T}^1}(x) \sim \frac{1}{2}\operatorname{Li}(x^d)$, which means that $P \cap \underline{T}^1$ has density $\frac{1}{2}$, and by Proposition 9.8 this implies that $P$ itself has density $\frac{1}{2}$.

## 9.3. The Chebotarev density theorem

### 9.3.1. Notation

9.3.1.1. We keep the hypotheses of §9.1.2 and §9.2 about $T$, and we also assume that $T$ is affine and normal, so that $T = \operatorname{Spec}\Lambda$ and $\underline{T} = \operatorname{Max}\Lambda$, where $\Lambda$ is an integrally closed domain, that is finitely generated as a $\mathbf{Z}$-algebra. Let $K$ be the field of fractions of $\Lambda$; it is a finitely generated extension of $\mathbf{Q}$ of transcendence degree $d-1$; conversely, every such extension is the field of fractions of a suitable $\Lambda$ of dimension $d$.

9.3.1.2. Let $K'$ be a finite Galois extension of $K$, with Galois group $G$. Denote by $\Lambda'$ the integral closure of $\Lambda$ in $K'$; it is integrally closed and its field of fractions is $K'$; moreover, it is stable under the action of $G$. Let $T'$ be $\operatorname{Spec}\Lambda'$; the group $G$ acts on $T'$, and we have a natural isomorphism $T'/G \simeq T$ which induces a bijection $\underline{T}'/G \simeq \underline{T}$.

9.3.1.3. *Decomposition, inertia and Frobenius.* Let $t'$ be a point of $\underline{T}'$ and let $t$ be its image in $\underline{T}$; they correspond to maximal ideals $\mathfrak{m}'$ and $\mathfrak{m}$ of $\Lambda'$ and $\Lambda$. As in §3.2.2, we may define :

• the *decomposition subgroup* $D_{t'}$, i.e. the subgroup of $G$ fixing $t'$;
• the *inertia subgroup* $I_{t'}$, i.e. the kernel of $D_{t'} \to \mathrm{Gal}(\kappa(t')/\kappa(t))$.

Since the map $D_{t'}/I_{t'} \to \mathrm{Gal}(\kappa(t')/\kappa(t))$ is an isomorphism (see e.g. [AC V-VI, Chap.V, §2, n° 2]), we thus get a canonical generator $\sigma_{t'/t}$ of $D_{t'}/I_{t'}$, which is the (arithmetic) *Frobenius* relative to $t'$. When $I_{t'} = 1$, then $T' \to T$ is étale above $t$, and we say that the covering $T' \to T$ is *unramified at* $t$; the element $\sigma_{t'/t}$ may then be viewed as an element of $D_{t'}$ and hence of $G$. Its conjugacy class depends only on $t$, and will be denoted by $\sigma_t$.

### 9.3.2. Statement of the density theorem

It is essentially the same as the standard one (which corresponds to the case $d = 1$, where $d = \dim T = \dim T'$), namely :

**Theorem 9.11.** *Let $C \subset G$ be stable under inner conjugation, and let $T(C)$ be the set of $t \in \underline{T}$ such that $T' \to T$ is unramified at $t$ and $\sigma_t$ belongs to $C$. Then there exists $c > 0$ such that*

$$\pi_{T(C)}(x) = \frac{|C|}{|G|}\mathrm{Li}(x^d) + O(x^d \exp(-c\sqrt{\log x})) \qquad \textit{for } x \to \infty.$$

The proof will be given in §9.4; it is a generalization of that of Theorem 9.1 (which is the special case where $G = 1$). It will also show that, if GRH is true, the error term can be replaced by $O(x^{d-\frac{1}{2}} \log x)$.

**Corollary 9.12.** *The set $T(C)$ has a density in the sense of* §9.2.1; *that density is equal to* $\frac{|C|}{|G|}$.

*Remark.* As in the number field case, this corollary can be interpreted as an *equidistribution statement.* More precisely, if we assume that $T'$ is étale over $T$ (i.e. that $G$ acts freely), and if we choose a numbering $\{t_1, ..., t_n, ...\}$ of $\underline{T}$ as in §9.2.1, then Corollary 9.12 says that the sequence of the Frobenius elements $\{\sigma_{t_1}, ..., \sigma_{t_n}, ...\}$ is equidistributed in $\mathrm{Cl}\, G$ for the Haar measure.

Note also the following useful consequence of Corollary 9.12, combined with Proposition 9.6 :

**Corollary 9.13.** *If $C \neq \varnothing$, then $T(C)$ is Zariski-dense in $T$.*

*Exercise* (A $\underline{T}$-analogue of the arithmetic progression theorem)

Let $n$ be an integer $\geqslant 1$ and let $\chi_n : \mathrm{Gal}(\overline{K}/K) \to (\mathbf{Z}/n\mathbf{Z})^\times$ be the $n$-th cyclotomic character. Let $m$ be the order of $\mathrm{Im}(\chi_n) \subset (\mathbf{Z}/n\mathbf{Z})^\times$. Let $a$ be an element of $(\mathbf{Z}/n\mathbf{Z})^\times$, and let $\underline{T}_a$ be the set of all $t \in \underline{T}$ such that $|t| \equiv a \pmod{n}$.

Prove :

i) $\underline{T}_a = \varnothing$ if $a \notin \mathrm{Im}(\chi_n)$;

ii) If $a \in \mathrm{Im}(\chi_n)$, $\underline{T}_a$ is frobenian with density $1/m$.

[Hint. Let $z_n$ be a primitive $n$-th root of unity in $\overline{K}$. The group $G = \mathrm{Im}(\chi_n)$ is the Galois group of the extension $K(z_n)/K$. Apply Corollary 9.12 to the pair $(\Lambda[z_n, 1/n], \Lambda[1/n])$ and observe that, if $t$ belongs to $\mathrm{Max}\,\Lambda[1/n]$, the corresponding Frobenius $\sigma_t$ is the image of $|t|$ in $(\mathbf{Z}/n\mathbf{Z})^\times$.]

### 9.3.3. Frobenian maps and frobenian sets

The definitions and results of §3.3 extend to $\underline{T}$ without any other change than replacing the finite sets $S$ of §3.3 by the subsets of $\underline{T}$ that are not Zariski-dense. For instance :

Let $\Omega$ be a set, let $S$ be a subset of $\underline{T}$ that is not Zariski-dense, and let $f : \underline{T} - S \to \Omega$ be a map. We say that $f$ is *S-frobenian* if there exists a finite Galois extension $K'/K$, and a map $\varphi : G \to \Omega$, where $G = \mathrm{Gal}(K'/K)$, having the following properties :

a) $\varphi$ is invariant under $G$-conjugation, i.e. factors through $G \to \mathrm{Cl}\,G$.

b) The extension $K'/K$ is " unramified outside $S$ ", i.e., if $\Lambda'$ is the integral closure of $\Lambda$ in $K'$, then $\Lambda'$ is étale over $\Lambda$ above every point of $\underline{T} - S$.

c) $f(t) = \varphi(\sigma_t)$ for every $t \in \underline{T} - S$.

A subset $\Sigma$ of $\underline{T}$ is called *S-frobenian* if its characteristic function is $S$-frobenian. When it is so, Theorem 9.11 shows that the density of $\Sigma$ exists and is a rational number ; when $\Sigma \neq \varnothing$, we have $\mathrm{dens}(\Sigma) > 0$.

Other definitions extend similarly. For instance, a subset $\Sigma$ of $\underline{T}$ is called *frobenian* if there exists a non-Zariski-dense $S$ such that $\Sigma - S \cap \Sigma$ is $S$-frobenian. We leave to the reader the task of stating and proving the analogues of Propositions 3.7, 3.8 and 3.9 in the $T$-context.

## 9.4. Proof of the density theorem

### 9.4.1. Strategy

We shall prove Theorem 9.11 by the usual method in such questions, that is :

a) Reduction to the case $d = 1$ , where $K$ is a number field ; this means applying algebraic geometry (especially the Deligne-Weil bounds) to the fibers of $T \to \mathrm{Spec}\,\mathbf{Z}$.

b) When $d = 1$, using standard results of analytic number theory.

Since we already have recalled in §3.2 the basic tools for b), we only have to take care of a). This will be done by introducing a 1-dimensional covering $T'_0 \to T_0$ with a commutative diagram :

$$\begin{array}{ccccc} & & T' & & \\ & \swarrow & & \searrow & \\ T & & & & T'_0 \\ & \searrow & & \swarrow & \\ & & T_0 & & \end{array}$$

with the property that the fibers of $T' \to T'_0$ and $T \to T_0$ are geometrically irreducible of dimension $\delta = d - 1$ [this is a kind of *Stein factorization* of $T' \to \operatorname{Spec} \mathbf{Z}$ and $T \to \operatorname{Spec} \mathbf{Z}$]. The reduction process consists in showing that Theorem 9.11 for $T' \to T$ follows by a $\delta$-shift from Theorem 9.11 for $T'_0 \to T_0$, which itself is equivalent to the standard Chebotarev Theorem given in §3.2.

### 9.4.2. Construction of $T'_0 \to T_0$

Let $K_0$ and $K'_0$ be the "fields of constants" of the function fields $K$ and $K'$, i.e. the largest number fields contained in $K$ and $K'$, cf. §9.1.7. The group $G$ acts on $K'_0$ and its fixed subfield is $K_0$ ; hence $K'_0$ is a Galois extension of $K_0$, and we have $\operatorname{Gal}(K'_0/K_0) = G/N$ where $N$ is the kernel of $G \to \operatorname{Aut}(K'_0)$.

We thus have a diagram of Galois extensions :

$$\begin{array}{ccc} K'_0 & \rightarrow & K' \\ {\scriptstyle G/N} \uparrow & & \uparrow {\scriptstyle G} \\ K_0 & \rightarrow & K. \end{array}$$

Since the extensions $K/K_0$ and $K'/K'_0$ are regular ([A IV-VII, Chap.5, §17, n° 4]), and of transcendence degree $\delta$, they correspond to geometrically irreducible varieties of dimension $\delta$. More precisely, after replacing $\underline{T}$ and $\underline{T}'$ by small enough affine open subschemes, we may assume that we have a diagram of schemes such as

$$\begin{array}{ccc} T' & \rightarrow & T'_0 \\ {\scriptstyle G} \downarrow & & \downarrow {\scriptstyle G/N} \\ T & \rightarrow & T_0 \end{array}$$

where $T_0$ and $T'_0$ are normal irreducible of dimension 1, with function fields $K_0$ and $K'_0$ respectively, the fibers of $T \to T_0$ and of $T' \to T'_0$ being geometrically irreducible [1] of dimension $\delta$. By taking even smaller open sets, we

[1] We are using here the fact that geometrical irreducibility, if true for a generic fiber, is true for every fiber in a dense open set, cf. [EGA IV, Théorème 9.7.7].

may also ensure that $G$ acts freely on $T'$ and that $G/N$ acts freely on $T'_0$, the quotients being $T$ and $T_0$ respectively. We can also assume that the ring $\Lambda$ (the affine ring of $T$) contains $1/\ell$ for at least one prime $\ell$. Such changes are allowed because they only modify $\pi_{T(C)}(x)$ by $O(x^{d-1}/\log x)$, which is smaller than the error term $O(x^d \exp(-c\sqrt{\log x}))$ of Theorem 9.11.

### 9.4.3. Reformulation of the theorem

It will be convenient to restate Theorem 9.11 in terms of a class function $f$ on $G$.

If we put

$$A_T(f,x) = \sum_{|t| \leqslant x} f(\sigma_t),$$

then Theorem 9.11 is equivalent to :

**Theorem 9.14.** *We have*

$$A_T(f,x) \;=\; \langle f,1\rangle_G \operatorname{Li}(x^d) + O(x^d \exp(-c\sqrt{\log x})) \quad \textit{for} \quad x \to \infty.$$

Indeed, by linearity, this statement is true if and only if it is so when $f$ is the characteristic function of a conjugacy class of $G$, in which case it is the same as Theorem 9.11.

### 9.4.4. The reduction process

Let $f$ be as above, and let $f^N$ be the corresponding function on $G/N$, cf. §5.1.4 and §9.3.3. Note that $\langle f,1\rangle_G = \langle f^N,1\rangle_{G/N}$. We may apply the definitions and notation of the previous section to the $G/N$-covering $T'_0 \to T_0$. Hence $f^N(\sigma_{t_0})$ makes sense for every $t_0 \in \underline{T}_0$ and so does $A_{T_0}(f^N,x)$.

Let us now choose a prime number $\ell$ such that the $\ell$-fiber of $T \to \operatorname{Spec}\mathbf{Z}$ is empty; this is possible by assumption, cf. §9.4.3. If $t_0$ is a point of $\underline{T}_0$, denote by $B(t_0)$ the sum of the $\ell$-adic Betti numbers (for the cohomology with proper support) of the $t_0$-fiber of $T' \to T_0$. The integers $B(t_0)$ remain bounded when $t_0$ varies, cf. e.g. Theorem 4.12; let $B$ be their upper bound.

**Proposition 9.15.** *For every $t_0 \in \underline{T}_0$, we have*

$$\left| \sum_{t \to t_0} f(\sigma_t) \;-\; |t_0|^{\delta} f^N(\sigma_{t_0}) \right| \leqslant (B-1).||f||.|t_0|^{\delta-\frac{1}{2}},$$

*where the sum extends to all $t \in \underline{T}$ above $t_0$ with the same degree as $t_0$, and $||f|| = \sum_{g\in G} |f(g)|$.*

The proof will be given in §9.4.6 and §9.4.7.

### 9.4.5. Proof of Theorems 9.11 and 9.14

Let us show how Theorem 9.14 (and hence Theorem 9.11) follows from Proposition 9.15. Let $A_T^1(f,x)$ be the same sum $\sum_{|t|\leqslant x} f(\sigma_t)$ as for $A_T(f,x)$, except that the summation is restricted to the $t$'s that are of degree 1. We have

(9.4.5.1) $A_T^1(f,x) = A_T(f,x) + O(x^{d-\frac{1}{2}})$, since the number of the $t$'s of degree $> 1$ is $O(x^{d-\frac{1}{2}})$, cf. Lemma 9.3. It will be then be enough to prove Theorem 9.14 with $A_T(f,x)$ replaced by $A_T^1(f,x)$.

We may rewrite $A_T^1(f,x)$ as :

$$A_T^1(f,x) = \sum_{|t_0|\leqslant x}^{1} \sum_{t\to t_0}^{1} f(\sigma_t),$$

where the first $\sum^1$ means that the summation is restrited to the $t_0 \in \underline{T}_0$ of degree 1, and the second one is similarly restricted to the $t$'s of degree 1. By Proposition 9.15, this gives :

$$A_T^1(f,x) = \sum_{|t_0|\leqslant x}^{1} |t_0|^\delta f^N(\sigma_{t_0}) + O(\sum_{|t_0|\leqslant x}^{1} |t_0|^{\delta-\frac{1}{2}}),$$

hence :

(9.4.5.2) $A_T^1(f,x) = \sum_{|t_0|\leqslant x}^{1} |t_0|^\delta f^N(\sigma_{t_0}) + O(x^{d-\frac{1}{2}})$.

By Lemma 9.3 applied to $T_0$, the number of $t_0$'s of degree $> 1$ with norm $\leqslant x$ is $O(x^{\frac{1}{2}})$. If such $t_0$'s were included in the above sums, their contribution would be $O(x^{d-\frac{1}{2}})$, hence would be negligible. This shows that

$${}^1_T(f,x) = \sum_{|t_0|\leqslant x} |t_0|^\delta f^N(\sigma_{t_0}) + O(x^{d-\frac{1}{2}}),$$

with no restriction on the degree of $t_0$. But $\underline{T}_0$ is the same as the set of non-zero prime ideals of the ring of integers of $K_0$, except that a finite number of such primes have been deleted. We may thus apply Theorem 3.6 to the Galois extension $K_0'/K_0$ , and we get :

$$\sum_{|t_0|\leqslant x} |t_0|^\delta f(\sigma_{t_0}) = \langle f^N, 1\rangle_{G/N} \operatorname{Li}(x^d) + O(x^d \exp(-c\sqrt{\log x})).$$

By applying (9.4.5.1) and (9.4.5.2), we get the same estimate for $A_T^1(f,x)$ and $A_T(f,x)$; since $\langle f^N, 1\rangle_{G/N} = \langle f, 1\rangle_G$, this concludes the proof of Theorem 9.14. Moreover, if one assumes that GRH holds for $K_0'$, the error term can be replaced by $O(x^{d-\frac{1}{2}} \log x)$.

### 9.4.6. Proof of Proposition 9.15 : rewriting $\sum f(\sigma_t)$ in terms of number of fixed points

*Notation.* If $\psi$ is an endomorphism of a $k$-variety $V$, we denote by $\mathrm{Fix}(\psi)$ the number of fixed points of $\varphi$ acting on $V(\overline{k})$ ; in all the cases we shall consider, that number is finite.

[A traditional notation for $\mathrm{Fix}(\psi)$ is $\Lambda(\psi)$, in honor of Lefschetz ; unfortunately, we are already using the letter $\Lambda$ for something else.]

We apply this notation to the case where $k = \kappa(t_0)$, where $t_0$ is a given point of $\underline{T}_0$, and $V$ is the fiber of $t_0$ for the map $T' \to T_0$. The group $G$ acts on $V$, and so does the Frobenius endomorphism $F$ relative to the finite field $\kappa(t_0)$. These two actions commute. For every $g \in G$, it is well known that the endomorphism $gF$ of $V$ has only a finite number of fixed points, so that $\mathrm{Fix}(gF)$ is a well defined positive integer.

The following lemma is standard in the theory of $L$-functions ; it relates the classical definition of these functions with their interpretation in terms of fixed points :

**Lemma 9.16.** *With the notation $f, t_0, \ldots$ as above, we have*

$$\sum_{t \to t_0} f(\sigma_t) = \frac{1}{|G|} \sum_{g \in G} f(g)\, \mathrm{Fix}(g^{-1}F),$$

*where the left sum is over the $t \in \underline{T}$ over $t_0$ with the same degree as $t_0$.*

*Proof.* Let us simplify the notation by writing :

$k = \kappa(t_0)$ ;

$X =$ fiber of $t_0$ in the projection $T \to T_0$ ;

$X' = T'(t_0) =$ fiber of $t_0$ in the projection $T' \to T_0$.

The group $G$ acts freely on $X'$ and we have $X'/G = X$. For every $g \in G$, denote by $X'_g$ the subset of $X'(\overline{k})$ made up of the fixed points of $g^{-1}F$. We have $|X'_g| = \mathrm{Fix}(g^{-1}F)$. Since $G$ acts freely, these sets are disjoint. A point $t \in X(\overline{k})$ is $k$-rational if and only if it belongs to the image by $X' \to X$ of one of the $X'_g$, in which case we have $\sigma_t = g$ in $\mathrm{Cl}\, G$ [2].

By linearity, it is enough to prove Lemma 9.16 when $f$ is the characteristic function $\varphi_C$ of a conjugacy class $C$ of $G$. Let us then define $X_C$ as the set of $t \in X(k)$ with $\sigma_t \in C$, and define $X'_C$ as the disjoint union of the $X'_g$ for all $g \in C$. The arguments above show that the inverse image of $X_C$ in $X'(\overline{k})$ is equal to $X'_C$ . Hence we have $|X_C| = \frac{1}{|G|}|X'_C|$ , which is equivalent to the formula of Lemma 9.16 for $f = \varphi_C$.

[2] This formula is correct if we make $G$ act on $T'$ and $X'$ by "transport de structure", which is a left action ; if we had used functoriality, which gives a right action, we would have $\sigma_t = g^{-1}$.

### 9.4.7. End of the proof of Proposition 9.15

Let us keep the notation of §9.4.6, and denote by $\mathrm{Fix}_0(g^{-1}F)$ the number of fixed points of $g^{-1}F$ acting on the $t_0$-fiber of $T_0' \to T_0$.

**Lemma 9.17.** *For every $g \in G$ we have :*

$$|\mathrm{Fix}(g^{-1}F) - |t_0|^{\delta}\,\mathrm{Fix}_0(g^{-1}F)| \leqslant |G|(B-1)|t_0|^{\delta-\frac{1}{2}}.$$

*Proof.*

Consider first the case $g = 1$. Then $\mathrm{Fix}(g^{-1}F) = \mathrm{Fix}(F)$ is the number of $k$-points of $X'(t_0)$, and we have a similar result for $\mathrm{Fix}_0(F)$, with $X'$ replaced by $X_0'$. The fibers of $X'(t_0) \to X_0'(t_0)$ are geometrically irreducible of dimension $\delta$, and the sum of their Betti numbers is $\leqslant B$.

If $\nu$ is the number of $k$-points of such a fiber, Deligne estimates imply that $|\nu - |t_0|^{\delta}| \leqslant (B-1)|t_0|^{\delta-\frac{1}{2}}$. By adding up, this shows that

$$|\mathrm{Fix}(F) \;-\; |t_0|^{\delta}\,\mathrm{Fix}_0(F)| \leqslant (B-1)|t_0|^{\delta-\frac{1}{2}}\mathrm{Fix}_0(F),$$

and since $|\mathrm{Fix}_0(F)| = |X_0'(t_0)| \leqslant (G:N)$ we get the bound we want.

The case $g \neq 1$ is analogous. One changes by a Galois twist the $k$-structures of $X'(t_0)$ and $X_0'(t_0)$ in such a way that the new Frobenius endomorphism is $g^{-1}F$; the Betti numbers do not change and the previous argument applies.

*Proof of Proposition 9.15.*

We want to prove the upper bound

$$(*) \quad |\sum_{t \to t_0} f(\sigma_t) \;-\; |t_0|^{\delta} f^N(\sigma_{t_0})| \leqslant (B-1).||f||.|t_0|^{\delta-\frac{1}{2}}.$$

Using Lemma 9.16, we may replace $\sum_{t\to t_0} f(\sigma_t)$ by

$$\frac{1}{|G|}\sum_{g\in G} f(g)\,\mathrm{Fix}(g^{-1}F),$$

where the fixed point number is relative to the action of $g^{-1}F$ on $T'(t_0)$. By the same lemma, applied to $T_0' \to T_0$, we may replace the term $|t_0|^{\delta} f^N(\sigma_{t_0})$ by $\frac{1}{|G/N|}\sum_{\gamma\in G/N} f^N(\gamma)\,\mathrm{Fix}_0(\gamma^{-1}F)$, which is equal to

$$\frac{1}{|G|}\sum_{g\in G} f(g)\,\mathrm{Fix}_0(g^{-1}F).$$

The left side of $(*)$ can thus be majorized by

$$\frac{1}{|G|}|\sum_{g\in G} |f(g)|.|\mathrm{Fix}(g^{-1}F) - |t_0|^{\delta}\,\mathrm{Fix}_0(g^{-1}F)|,$$

and by applying Lemma 9.17 to each term $|\mathrm{Fix}(g^{-1}F) - |t_0|^{\delta}\mathrm{Fix}_0(g^{-1}F)|$, we get $(*)$.

This concludes the proof of Proposition 9.15, and hence of Theorems 9.11 and 9.14.

## 9.5. Relative schemes

### 9.5.1. The relative setting

We keep the notation of §9.3.1 : the affine scheme $T = \mathrm{Spec}\ \Lambda$ is normal, irreducible, of dimension $d$ and the morphism $T \to \mathrm{Spec}\ \mathbf{Z}$ is dominant ; its field of fractions $K$ is a finitely generated extension of $\mathbf{Q}$, of transcendence degree $d-1$.

Let $X \to T$ be a scheme over $T$, that is of finite type (over $T$ or over $\mathrm{Spec}\ \mathbf{Z}$, it amounts to the same). Denote by $X_0$ its generic fiber ; it is a $K$-variety. For every $t \in \underline{T}$, let $X_t$ denote the fiber of $X$ over $t$ ; it is a scheme over the residue field $\kappa(t)$. If $e$ is an integer $> 0$, we denote by $N_X(t^e)$ the number of points of $X_t$ with values in a field extension of $\kappa(t)$ of degree $e$ [we could also accept any $e \in \mathbf{Z}$, as in §1.5] ; when $e = 1$, we write $N_X(t)$ instead of $N_X(t^e)$.

### 9.5.2. A claim

We only offer the following admittedly imprecise statement :

**Claim 9.18.** *Almost all the definitions, results and conjectures given in Chapters 6, 7 and 8 for the function $p \mapsto N_X(p)$ can be extended to the function $t \mapsto N_X(t)$.*

[Note that this includes in particular the case where $K$ is a number field.]

The justification of the claim is that the arguments of Chapters 6, 7 and 8 only use :

(a) basic facts on schemes over finite fields ;

(b) general facts on group representations and densities ;

(c) the Chebotarev density theorem ;

(d) the frobenian interpretation of $N_X(p^e)$.

Items (a) and (b) have already been handled in Chap.4 and Chap.5, and (c) has been done in §9.3 and §9.4. As for (d), see §9.5.3 below.

### 9.5.3. The $\ell$-adic representations

We need to replace the Galois group $\Gamma_S$ of §6.1 by the fundamental group $\pi_1(T)$ of $T$ relative to the geometric point $\mathrm{Spec}\,\overline{K} \to T$, cf. e.g. [SGA 1, V, §§7-8]. The field-theoretic definition of $\pi_1(T)$ is that it is the Galois group of $K_T/K$, where $K_T$ is the maximal subextension of $\overline{K}$ that is unramified over $T$ [3].

Assume for simplicity that $X$ is separated (the general case can be handled by the same technique as in §6.1), and let us denote by $\overline{X}$ the geometric generic fiber of $X \to T$, i.e. the $\overline{K}$-variety deduced from $X_0$ by the base change $K \to \overline{K}$. Let $H^i_K(X, \ell)$ be the $i$-th cohomology group of $\overline{X}$ with coefficients in $\mathbf{Q}_\ell$. There is a natural action of $\Gamma_K$ on this $\mathbf{Q}_\ell$-vector space.

**Theorem 9.19.** *There exists a closed non-Zariski-dense subscheme $S$ of $T$ with the following two properties :*

1) *For every $i$, the action of $\Gamma_K$ on $H^i_K(X, \ell)$ is unramified over $T - S$, i.e. factors through the group $\pi_1(T - S)$.*

2) *For every closed point $t$ of $T - S$, and every integer $e$, we have*

$$N_X(t^e) = \sum_i (-1)^i \mathrm{Tr}(g_t^e | H^i(X, \ell)),$$

*where $g_t$ denotes the geometric Frobenius of $t$, viewed as a conjugacy class in $\pi_1(T - S)$.*

The proof is the same as that of Theorem 6.1.
The properties of the $\ell$-adic representations $H^i(X, \ell)$ given in Chap.6 and Chap.7 extend without any change. One only has to replace "finite set $S$ of primes" by "closed non-Zariski-dense subscheme $S$ of $T$ ". Let us for instance write down the analogue of Theorem 6.14 :

**Theorem 9.20.** *Let $X$ and $Y$ be two $T$-schemes of finite type over $T$. Suppose that $|N_X(t) - N_Y(t)|$ remains bounded when $t$ varies. Then there exists a closed non-Zariski-dense subscheme $S$ of $T$ such that the map $t \mapsto N_X(t) - N_Y(t)$ is $S$-frobenian, in the sense defined in §9.3.3.*
The proof is the same.

We leave to the reader the task of translating most of the other results of Chap.6 and Chap.7 in a similar way.

### 9.5.4. Sato-Tate

As for Chap.8 (Sato-Tate), there is not much to change either. Note that the $\mu$-equidistribution conjecture should be formulated in terms of a *numbering* of $\underline{T}$, as in §9.2.1. Or, equivalently, it could be stated as

$$\mu(\varphi) = \lim_{x \to \infty} \frac{1}{\pi_T(x)} \sum_{|t| \leqslant x} \varphi(f(t)),$$

with the same notation as in §8.1.2.

[3] More concretely, a finite subextension $K'/K$ is contained in $K_T$ if and only if the normalization of $\Lambda$ in $K'$ is étale over $\Lambda$.

There is one change that is worth mentioning : the existence of the element $\gamma$ in 8.2.3.4 should only be postulated when $T$ has a smooth $\mathbf{R}$-point. This makes the list of the possible Sato-Tate groups a bit larger (up to 3 for elliptic curves, and up to about 50 for curves of genus 2, see [FKRS 11] ).

An interesting fact is that the Sato-Tate conjecture is sometimes *easier to prove* in the higher dimensional case ($d > 1$) than in the number field case, thanks to the information given by the geometric monodromy (as done by Deligne in characteristic $p$, cf. [De 80]). An early instance of this is [Bi 68] [4] which handles the case of the elliptic curve $y^2 = x^3 - ax - b$ over $T = \operatorname{Spec} \mathbf{Z}[a, b]$, where $a$ and $b$ are independent indeterminates.

[4] Beware of a misprint in that paper : the values of $S_1(p), S_2(p), \ldots$ given in Theorem 2 should be multiplied by $p - 1$.

# REFERENCES

[A IV-VII] N. Bourbaki , *Algèbre - Chapitres* 4-7, Paris, Masson, 1981 ; new printing, Springer-Verlag, 2006 ; English translation, *Algebra II*, Springer-Verlag, 1989.

[A VIII] ——— *Algèbre - Chapitre* 8 : *Anneaux et Modules Semi-simples*, new revised edition, Springer-Verlag, 2011 ; English translation of the first edition, *Algebra*, Springer-Verlag, 1998.

[AC V-VI] ——— *Algèbre Commutative - Chapitre* 5 : *Entiers - Chapitre* 6 : *Valuations*, Paris, Hermann, 1975 ; new printing, Springer-Verlag, 2006 ; English translation, *Commutative Algebra*, Springer-Verlag, 1989.

[AP 08] J. Aguirre & J.C. Peral, *The trace problem for totally positive algebraic integers*, in *Number Theory and Polynomials*, edit. J. McKee & C. Smyth, LMS Lect. Notes 352, Cambridge, 2008, 1-19.

[Ar 23] E. Artin, *Über eine neue Art von L-Reihen*, Hamb. Abh. 3 (1923), 89-108 (= *Coll. Papers*, n° 3).

[Ar 02] J. Arthur, *A note on the automorphic Langlands group*, Canad. Math. Bull. 45 (2002), 466-482.

[ATLAS] J.H. Conway, R.T. Curtis, S.P. Norton, R.A. Parker & R.A. Wilson, *Atlas of Finite Groups*, Clarendon Press, Oxford, 1985 ; second corrected edition, 2003.

[Ax 67] J. Ax, *Solving diophantine problems modulo every prime*, Ann. Math. 87 (1967), 161-183.

[BLGHT 11] T. Barnet-Lamb, D. Geraghty, M. Harris & R. Taylor, *A family of Calabi-Yau varieties and potential automorphy II*, Publ. Res. Inst. Math. Sci. 47 (2011), 29-98.

[BCDT 01] C. Breuil, B. Conrad, F. Diamond & R. Taylor, *On the modularity of elliptic curves over* $\mathbf{Q}$ : *wild* 3*-adic exercises*, J. A. M. S. 14 (2001), 843-939.

[BCR 98] J. Bochnak, M. Coste & M-F. Roy, *Real Algebraic Geometry*, Ergebn. der Math. (3) 36, Springer-Verlag, 1998.

[Bi 68] B.J. Birch, *How the number of points of an elliptic curve over a fixed prime field varies*, J. London Math. Soc. 43 (1968), 57-60.

[Bl 52] F. van den Blij, *Binary forms of discriminant* $-23$, Indag. Math. 14 (1952), 498-503.

[BLR 90] S. Bosch, W. Lütkebohmert & M. Raynaud, *Néron Models*, Ergebn. der Math. (3) 21, Springer-Verlag, 1990.

[Bo 80] F.A. Bogomolov, *Sur l'algébricité des représentations $\ell$-adiques*, C.R.A.S. 290 (1980), 701-703.

[Ca 07] H. Carayol, *La conjecture de Sato-Tate*, Séminaire Bourbaki 2006/207, n° 977 ; Astérisque 317 (2008), 345-391.

[CC 92] P.J. Cameron & A.M. Cohen, *On the number of fixed point free elements in a permutation group*, Discrete Math. 106/107 (1992), 135-138.

[CE 56] H. Cartan & S. Eilenberg, *Homological Algebra*, Princeton Univ. Press, Princeton, 1956.

[CE 11] J-M. Couveignes & B. Edixhoven (edit.), *Computational Aspects of Modular Forms and Galois Representations*, Ann. Math. Studies 176, Princeton Univ. Press, Princeton, 2011.

[Ch 25] N.G. Tschebotareff (a.k.a. Chebotarev, Chebotarëv or Chebotaryov), *Die Bestimmung der Dichtigkeit einer Menge von Primzahlen, welche zu einer gegebenen Substitutionsklasse gehören*, Math. Ann. 95 (1925), 191-228.

[CHT 08] L. Clozel, M. Harris & R. Taylor, *Automorphy for some $\ell$-adic lifts of automorphic mod $\ell$ Galois representations*, Publ. Math. IHÉS 108 (2008), 1-181.

[Cl 08] L. Clozel, *The Sato-Tate conjecture*, in *Current Developments in Mathematics (2006)*, Int. Press Boston, 2008, 1-34.

[CL 06] A. Chambert-Loir, *Compter (rapidement) le nombre de solutions d'équations sur les corps finis*, Séminaire Bourbaki 2006/2007, n° 968 ; Astérisque 317 (2008), 39-90.

[Co 93] H. Cohen, *A Course in Computational Number Theory*, Springer-Verlag, 1993.

[Co 97] J.H. Conway, *The sensual (quadratic) form*, Carus Math. Monograph 26, The Math. Association America, Washington, 1997.

[Cox 89] D.A. Cox, *Primes of the Form $x^2 + ny^2$*, John Wiley & Sons, New York, 1989.

[Cr 97] J.E. Cremona, *Algorithms for Modular Elliptic Curves*, second edition, Cambridge U. Press, Cambridge, 1997.

[CS 03] P. Colmez & J-P. Serre (edit.), *Grothendieck-Serre Correspondence*, bilingual edition, AMF & SMF, 2003.

[De 53] M. Deuring, *Die Zetafunktion einer algebraischen Kurve von Geschlechte Eins*, Nachr. Akad. Wiss. Göttingen (1953), 85-94.

[De 68] P. Deligne, *Formes modulaires et représentations $\ell$-adiques*, Séminaire Bourbaki 1968/1969, n° 355 ; Springer Lect. Notes 179 (1971), 139-172.

[De 74] ———, *La Conjecture de Weil*, Publ. Math. IHÉS 43 (1974), 273-308.

[De 80] ———, *La Conjecture de Weil II*, Publ. Math. IHÉS 52 (1980), 137-252.

[De 90] ———, *Catégories tannakiennes*, in *Grothendieck Festschrift*, vol. II, 111-195, Progress in Math. 87, Birkhäuser Boston, 1990.

[De 09] J-P. Demailly, *Complex Analytic and Differential Geometry*, Open-Content Book, Univ. Grenoble, 2009 ; available as [agbook.pdf].

[Di 39] G. Lejeune-Dirichlet, *Recherches sur quelques applications de l'analyse infinitésimale à la théorie des nombres*, J. Crelle 19 (1839), 324-369 and 21 (1840), 134-155 (= *Werke*, I, 413-496).

[Do 07] I.V. Dolgachev, *Topics in Classical Algebraic Geometry. Part I*, Lecture Notes, Univ. Michigan, 2007.

[Dr 91] L. van den Dries, *A remark on Ax's theorem on solvability modulo primes*, Math. Z. 208 (1991), 65-70.

[DS 74] P. Deligne & J-P. Serre, *Formes modulaires de poids* 1, Ann. Sci. E.N.S. 7 (1974), 507-530 (= J-P. Serre, *Coll. Papers*, vol. III, n° 101).

[Dw 60] B. Dwork, *On the rationality of the zeta function of an algebraic variety*, Amer. J. Math. 82 (1960), 631-648.

[EGA] A. Grothendieck, Éléments de Géométrie Algébrique (rédigés avec la collaboration de J. Dieudonné), Chap. 0-IV, Publ. Math. I.H.É.S 8, 11, 17, 20, 24, 28, 32 (1960-1967).

[EJ 10] A-S. Elsenhans & J. Jahnel, *The discriminant of a cubic surface*, arXiv : 1006.0721.

[El 87] N.D. Elkies, *The existence of infinitely many supersingular primes for every elliptic curve over* $\mathbf{Q}$, Invent. math. 89 (1987), 561-567.

[El 91] ———, *Distribution of supersingular primes*, in *Journées Arithmétiques de Luminy 1989*, Astérisque 198-199-200 (1991), 127-132.

[El 04] J.S. Ellenberg, $K3$ *surfaces over number fields with geometric Picard number one*, in *Arithmetic of higher-dimensional algebraic varieties*, Progr. Math. 226 (2004), Birkhäuser Boston, 135-140.

[Fa 84] G. Faltings, *Complements to Mordell*, Chapter VI of *Rational Points*, edit. G.Faltings, G. Wüstholz et al., Aspects of Mathematics E6, Friedr. Vieweg & Sohn, Braunschweig/Wiesbaden, 1984.

[Fa 88] ———, *p-adic Hodge theory*, J.A.M.S. 1 (1988), 255-299.

[FJ 08] M.D. Fried & M. Jarden, *Field Arithmetic*, Ergebn. der Math. 11 (3), third edition, revised by M. Jarden, Springer-Verlag, Heidelberg, 2008.

[FK 88] E. Freitag & R. Kiehl, *Étale Cohomology and the Weil Conjectures*, Ergebn. der Math. 13 (3), Springer-Verlag, Heidelberg, 1988.

[FKRS 11] F. Fité, K.S. Kedlaya, V. Rotger & A.V. Sutherland, *Sato-Tate distributions and Galois endomorphism modules in genus* 2, in preparation.

[FM 85] J-M. Fontaine & W. Messing, *p-adic periods and p-adic étale cohomology*, in *Current Trends in Arithmetical Algebraic Geometry*, AMS Contemp. Math. 67 (1985), 1987, 179-207

[Fr 28] R. Fricke, *Lehrbuch der Algebra III*, Vieweg & Sohn, Braunschweig, 1928.

[Fr 96] F.G. Frobenius, *Über Beziehungen zwischen den Primidealen eines algebraischen Körpers und den Substitutionen seiner Gruppe*, Sitz. Akad. Wiss. Berlin (1896), 689-703 (= *Ges. Abh.*, vol. II, n° 52).

[FRV] N. Bourbaki, *Variétés différentielles et analytiques - Fascicule de résultats §§8-15*, Hermann, Paris, 1971 ; new printing, Springer-Verlag, 2007.

[Ga 30] E. Galois, *Sur la théorie des nombres*, Bulletin de Férussac 13 (1830), 428-436 (= *Oeuvres Mathématiques*, J. Liouville 11 (1846), 398-407).

[Go 01] R. Godement, *Analyse Mathématique, vol. 4, Intégration et théorie spectrale, analyse harmonique, le jardin des délices modulaires*, Springer-Verlag, 2001.

[Gr 64] A. Grothendieck, *Formule de Lefschetz et rationalité des fonctions L*, Séminaire Bourbaki 1964/1965, exposé 279 ; reprinted in *Dix exposés sur la cohomologie des schémas*, North-Holland, Amsterdam (1968), 31-45.

[GS 96] H. Gillet & C. Soulé, *Descent, motives and K-theory*, J. Crelle 478 (1996), 127-176.

[GV 09] G. van der Geer & M. van der Vlugt, *Tables of curves with many points*, available on van der Geer's homepage ; see also www.manypoints.org.

[Ha 77] R. Hartshorne, *Algebraic Geometry*, Springer-Verlag, 1977.

[Ha 06] M. Harris, *The Sato-Tate conjecture : introduction to the proof*, preprint.

[Ha 09] ______ , *Galois representations, automorphic forms, and the Sato-Tate Conjecture*, Proceedings of the Clay Research Conferences, 2007-2008, to appear.

[He 20] E. Hecke, *Eine neue Art von Zetafunktionen und ihre Beziehungen zur Verteilung der Primzahlen. Zweite Mitteilung*, Math. Zeit. 6 (1920), 11-51 (= *Math. Werke*, n° 14).

[Hi 64] H. Hironaka, *Resolution of singularities of an algebraic variety over a field of characteristic zero*, Ann. Math. 79 (1964), 109-326.

[Ho 01] E.W. Howe, *Isogeny classes of abelian varieties with no principal polarization*, in *Moduli of Abelian Varieties*, edit. C. Faber, G. van der Geer & F. Oort, Progr. Math. 195, Birkäuser, Basel (2001), 203-216.

[HSBT 10] M. Harris, N. Shepherd-Barron & R.Taylor, *A family of Calabi-Yau varieties and potential automorphy*, Ann. Math. 171 (2010), 779-813.

[Il 10] L. Illusie, *Constructibilité générique et uniformité en $\ell$*, unpublished preprint, Orsay, 2010.

[INT] N. Bourbaki, *Intégration – Chapitres I - V*, two volumes, second edition, Hermann, Paris, 1967 ; new printing, Springer-Verlag, 2007 ; English translation, *Integration I*, Springer-Verlag, 2004.

[Ka 52] I. Kaplansky, *Modules over Dedekind rings and valuation rings*, T.A.M.S. 72 (1952), 327-340.

[Ka 94] N.M. Katz, *Review of $\ell$-adic cohomology*, AMS Symp. Pure Math. 55 (1994), Part 1, 21-30.

[Ka 01a] ______ , *Sums of Betti numbers in arbitrary characteristic*, Finite Fields and their Applications 7 (2001), 29-44.

[Ka 01b] ______ , *L-Functions and Monodromy : Four Lectures on Weil II*, Advances in Math. 160 (2001), 81-132.

[KL 86] N.M. Katz & G. Laumon, *Transformation de Fourier et majoration de sommes exponentielles*, Publ. Math. I.H.É.S 62 (1986), 361-418.

[Kn 81] D. Knuth, *The Art of Computer Programming*, volume II, *Seminumerical Algorithms*, Addison-Wesley, 1981 ; third edition, 1998.

[Ko 07] J. Kollár, *Lectures on Resolution of Singularities*, Ann. Math. Studies 166, Princeton Univ. Press, 2007.

[KP 99] J. Kaczorowki & A. Perelli, *The Selberg class : a survey*, in *Number Theory in Progress*, vol.1, Walter de Gruyter, New York (1999), 953-992

[KS 99] N.M. Katz & P. Sarnak, *Random Matrices, Frobenius Eigenvalues, and Monodromy*, AMS Colloquium Publ. 45 (1999).

[KS 08] K.S. Kedlaya & A.V. Sutherland, *Computing L-series of hyperelliptic curves*, Springer Lect. Notes in Computer Science 5011 (2008), 312-326.

[KS 09] ______ , *Hyperelliptic Curves, L-polynomials, and Random Matrices*, in *Arithmetic, Geometry, Cryptography and Coding Theory*, AMS Contemp. Math. 487 (2009), 119–162.

[La 67] R.P. Langlands, *Euler Products*, Lecture notes, Yale Univ., New Haven, 1967.

[La 79] ______ , *Automorphic representations, Shimura varieties, and motives. Ein Märchen*, in *Automorphic Forms, Representations and L-Functions*, AMS Proc. Symp. Pure Math, 33-II (1979), 205-246.

[La 80] S. LaMacchia, *Polynomials with Galois group* $\mathbf{PSL}(2,7)$, Comm. in Algebra 8 (1980), 983-992.

[La 81] G. Laumon, *Comparaison de caractéristiques d'Euler-Poincaré en cohomologie* $\mathbf{Q}_\ell$*-adique*, C.R.A.S. 292 (1981), 209-212.

[La 02] K. Lauter, *The maximum or minimum number of rational points on genus three curves over a finite field*, Compositio Math. 134 (2002), 87-111.

[Le 74] H.W. Lenstra, Jr., *Rational functions invariant under a finite abelian group*, Invent. math. 25 (1974), 299-325.

[LIE III] N. Bourbaki, *Groupes et Algèbres de Lie - Chapitre III - Groupes de Lie*, Paris, Hermann, 1972 ; new printing, Springer-Verlag, 2007 ; English translation, *Lie Groups and Lie Algebras*, Springer-Verlag, 1998.

[Li 95] R. Livné, *Motivic orthogonal two-dimensional representations of* $\mathrm{Gal}(\overline{\mathbf{Q}}/\mathbf{Q})$, Israel J. Math. 92 (1995), 149-156.

[LO 77] J.C. Lagarias & A.M. Odlyzko, *Effective versions of the Chebotarev density theorem*, in *Algebraic Number Fields*, edit. A. Fröhlich, Academic Press, London, 1977, 409-464.

[LS 96] H.W. Lenstra, Jr & P. Stevenhagen, *Chebotarëv and his density theorem*, Math. Intelligencer 18 (1996), 26-37.

[LT 76] S. Lang & H. Trotter, *Frobenius distributions in* $\mathbf{GL}_2$*-extensions*, Springer Lect. Notes 504 (1976).

[Man 86] Y.I. Manin, *Cubic Forms : Algebra, Geometry, Arithmetic*, second edition, North Holland, Amsterdam, 1986.

[Maz 78] B. Mazur, *Modular curves and the Eisenstein ideal*, Publ. Math. IHÉS, 47 (1978), 33-186.

[Mi 80] J.S. Milne, *Étale Cohomology*, Princeton Math. Series 32, Princeton, 1980.

[Mi 87] H. Minkowski, *Zur Theorie der positiven quadratischen Formen*, J. Crelle 101 (1887), 196-202 (= *Ges. Abh.*, Band I, n° VI).

[Mo 04] B. Moonen, *An introduction to Mumford-Tate groups*, unpublished lecture notes, Amsterdam, 2004

[Mu 66] D. Mumford, *Families of abelian varieties*, in *Algebraic Groups and Discontinuous Subgroups*, AMS Proc. Symp. Pure Math. IX (1966), 347-351.

[OEIS] N.J.A. Sloane, *The on-line encyclopedia of integer sequences*, available at *http : //oeis.org*.

[Pi 90] J. Pila, *Frobenius maps of abelian varieties and finding roots of unity in finite fields*, Math. Comp. 55 (1990), 745-763.

[Pi 97] R. Pink, *The Mumford-Tate conjecture for Drinfeld modules*, Publ. Res. Inst. Math. Sci. 33 (1997), 393-425.

[Pr 23] H. Prüfer, *Untersuchungen über die Zerlegbarkeit der abzählbaren abelschen Gruppen*, Math. Zeit. 17 (1923), 35-61.

[PTV 92] C. Peters, J. Top & M. van der Vlugt, *The Hasse zeta function of a K3-surface related to the number of words of weight 5 in the Melas codes*, J. Crelle 432 (1992), 151-176.

[Ra 03] E.M. Rains, *Images of eigenvalue distributions under power maps*, Probab. Theory Relat. Fields 125 (2003), 522-538.

[RS 94] M. Rubinstein & P. Sarnak, *Chebyshev's Bias*, Experimental Math. 3 (1994), 173-197.

[RZ 00] L. Ribes & P. Zalesskii, *Profinite Groups*, Ergebn. der Math. 40 (3), Springer-Verlag, Heidelberg, 2000.

[Sa 94] T. Saito, *Jacobi sum Hecke characters, de Rham discriminant, and the determinant of* $\ell$*-adic cohomologies*, J. Alg. Geom. 3 (1994), 411-434.

[SS 12] T. Saito & J-P. Serre, *The discriminant and the determinant of a hypersurface of even dimension*, to appear.

[Sc 85] R.J. Schoof, *Elliptic curves over finite fields and the computation of square roots mod p*, Math. Comp. 44 (1985), 483-494.

[Se 62] J-P. Serre, *Corps Locaux*, Hermann, Paris, 1962 ; fourth corrected edition, 2004 ; English translation, *Local Fields*, Springer-Verlag, 1979.

[Se 64] ———— , *Cohomologie Galoisienne*, Springer Lect. Notes 5 (1964) ; fifth revised edition, 1994 ; English translation, *Galois Cohomology*, Springer-Verlag, 1997.

[Se 65] ———, *Zeta and L functions*, in *Arithmetical Algebraic Geometry*, Harper and Row, New York, 1965, 82-92 (= *Coll. Papers*, vol. II, n° 64).

[Se 68] ———, *Abelian ℓ-adic Representations and Elliptic Curves*, W.A. Benjamin, 1968 ; AK Peters, 1998.

[Se 77] ———, *Représentations ℓ-adiques*, in *Kyoto Symposium on Algebraic Number Theory*, Japan Soc. for the Promotion of Science (1977), 177-193 (= *Coll. Papers*, vol. III, n° 112).

[Se 78] ———, *Représentations Linéaires des Groupes Finis*, third revised edition, Hermann, Paris, 1978 ; English translation, *Linear Representations of Finite Groups*, Springer-Verlag, 1977.

[Se 79] ———, *Groupes algébriques associés aux modules de Hodge-Tate*, Astérisque 65 (1979), 155-188 (= *Coll. Papers*, vol. III, n° 119).

[Se 81] ———, *Quelques applications du théorème de densité de Chebotarev*, Publ. Math. I.H.É.S 54 (1981), 123-201 (= *Coll. Papers*, vol. III, n° 125).

[Se 91] ———, *Lettres à Ken Ribet du 1/1/1981 et du 29/1/1981*, reproduced in *Coll. Papers*, vol. IV, n° 133.

[Se 93] ———, *Gèbres*, L'Ens. Math. 39 (1993), 33-85 (= *Coll. Papers*, vol. IV, n° 160).

[Se 94] ———, *Propriétés conjecturales des groupes de Galois motiviques et des représentations ℓ-adiques*, in *Motives*, AMS Proc. Symp. Pure Math. 55 (1994), vol.I, 377-400 (= *Coll. Papers*, vol. IV, n° 161).

[Se 98] ———, *Letter to C. Smyth, 2/24/1998*, reproduced in [AP 08], 14-18.

[Se 02] ———, *On a theorem of Jordan*, Math. Medley 29 (2002), 3-18 ; Bull. AMS 40 (2003), 429-440.

[Se 07] ———, *Bounds for the orders of the finite subgroups of $G(k)$*, in *Group Representation Theory*, edit. M. Geck, D. Testerman & J. Thévenaz, EPFL Press, Lausanne, 2007, 403-450.

[SGA 1] A. Grothendieck, *Revêtements étales et groupe fondamental*, Springer Lect. Notes 224 (1971) ; revised edition, *Documents Mathématiques* 3, SMF, 2003.

[SGA 3] M. Demazure & A. Grothendieck, *Schémas en Groupes*, 3 vol., Springer Lect. Notes 151, 152, 153 (1970) ; revised edition, *Documents Mathématiques*, SMF, to appear.

[SGA 4] M. Artin, A. Grothendieck & J-L. Verdier, *Théorie des Topos et Cohomologie Étale des Schémas*, 3 vol., Springer Lect. Notes 269, 270, 305 (1972-1973).

[SGA $4\frac{1}{2}$] P. Deligne, *Cohomologie Étale*, Springer Lect. Notes 569 (1977).

[SGA 5] A. Grothendieck, *Cohomologie ℓ-adique et Fonctions L*, edit. L. Illusie, Springer Lect. Notes 589 (1977).

[Sh 66] G. Shimura, *A reciprocity law in non-solvable extensions*, J. Crelle 221 (1966), 209-220 (= *Coll. Papers*, vol. I, [66a]).

[Sh 98] ———, *Abelian Varieties with Complex Multiplication and Modular Functions*, Princeton Univ. Press, Princeton, 1998.

[Si 45] C.L. Siegel, *The trace of totally positive and real algebraic integers*, Ann. Math. 46 (1945), 302-312 (= *Ges. Abh.*, vol. III, n° 48).

[SI 77] T. Shioda & H. Inose, *On singular $K3$ surfaces*, in *Complex Analysis and Algebraic Geometry*, ed. W.L. Baily & T. Shioda, Iwanami & Cambridge U. Press, 1977, 119-136.

[Sm 84] C. Smyth, *Totally positive algebraic integers of small trace*, Ann. Inst. Fourier 33 (1984), 1-28.

[St 99] R.P. Stanley, *Enumerative Combinatorics, volume 2*, Cambridge Univ. Press, Cambridge, 1999.

[Sw 69] R.G. Swan, *Invariant rational functions and a problem of Steenrod*, Invent. math. 7 (1969), 148-158.

[Sw 10] Sir Peter Swinnerton-Dyer, *Cubic surfaces over finite fields*, Math. Proc. Camb. Phil. Soc. 149 (2010), 385-388.

[Ta 65] J. Tate, *Algebraic cycles and poles of zeta functions*, in *Arithmetic Algebraic Geometry (Proc. Conf. Purdue Univ., 1963)*, Harper & Row, New York (1965), 93-110.

[Ta 69] ———, *Classes d'isogénie des variétés abéliennes sur un corps fini (d'après T.Honda)*, Séminaire Bourbaki n° 352, 1969 ; Springer Lect. Notes 179 (1971), 95-110.

[Ta 79] ———, *Number theoretic background*, in *Automorphic Forms, Representations and L-Functions*, AMS Proc. Symp. Pure Math, 33-II (1979), 3-26.

[Ta 08] R. Taylor, *Automorphy for some $\ell$-adic lifts of automorphic mod $\ell$ Galois representations* II, Publ. Math. IHÉS 108 (2008), 183-239.

[Te 85] T. Terasoma, *Complete intersections with middle Picard number* 1 *defined over* $\mathbf{Q}$, Math. Zeit. 189 (1985), 289-296.

[Vo 70] V.E. Voskresenskiĭ, *On the question of the structure of the subfield of invariants of a cyclic group of automorphisms of the field* $\mathbf{Q}(x_1, ..., x_n)$, (in Russian), Izv. Akad. Nauk SSSR 34 (1970), 366-375 ; English translation, Math. USSR Izv. 4 (1970), 371-380.

[We 34] A. Weil, *Une propriété caractéristique des groupes de substitutions linéaires*, C.R.A.S. 198 (1934), 1739-1742 & C.R.A.S. 199 (1934), 180-182 (= *Coll. Papers*, vol. I, [1934b] & [1934c]).

[We 48] ———, *Sur les courbes algébriques et les variétés qui s'en déduisent*, and *Variétés abéliennes et courbes algébriques*, Hermann, Paris, 1948 ; reprinted in 1971 under the joint title *Courbes algébriques et variétés abéliennes.*

[We 54] ———, *Abstract versus classical algebraic geometry*, Proc. I. C. M. Amsterdam (1954), vol. III, 550-558 (= *Coll. Papers*, vol. II, [1954h]).

[We 61] ———, *Adeles and algebraic groups*, with appendices by M. Demazure and T. Ono, I.A.S. Princeton, 1961 ; Birkhäuser, Boston, 1982.

[We 67] ———, *Über die Bestimmung Dirichletscher Reihen durch Funktionalgleichungen*, Math. Ann. 168 (1967), 149-156 (= *Coll. Papers*, vol. III, [1967a]).

[Wi 95] A. Wiles, *Modular elliptic curves and Fermat's last theorem*, Ann. Math. 141 (1995), 443-551.

[Za 02] Y.G. Zarhin, *Very simple 2-adic representations and hyperelliptic jacobians*, Moscow Math. J. 2 (2002), 403-431.

# Index of Notations


$\mathbf{A}^1$ = affine line : 6.1.1
$\mathbf{A}^n$ = affine space of dimension $n$ : 6.1.3
$A_m(C,x), A_m(f,x)$ : 3.2.4
$(A_0), \ldots, (A_5)$ : 8.2.2, 8.2.3
$B = B(X) + B(Y)$ : 6.3.3
$B$ = upper bound of the $B(t_0)$ : 9.4.4
$B_i, B_i(X)$ = Betti numbers with proper support : 4.5, 6.3.3
$B(X) = \sum B_i(X)$ : 6.3.3
$B^i(X)$ = virtual Betti numbers : 7.1.3
$C_{G,K}$ : 5.1.1
$\mathrm{Cl}\, G$ = set of conjugacy classes of $G$ : 3.2.2
$\mathrm{Cl}(G,K), \mathrm{Cl}(G,A)$ : 5.1.1
$C_n$ = $n$-th Catalan number : 8.1.5.2
$d$ = dimension of the scheme $T$ : 9.1.2
$\delta = d - 1$ = relative dimension of $T \to T_0$ : 9.4.1
$\mathrm{dens}^{\mathrm{haar}}$, $\mathrm{dens}^{\mathrm{zar}}$ : 5.2.1
$d(t)$ = degree of a closed point $t$ : 9.1.1
$D_w, D_{\overline{v}}$ : 3.2.1, 4.8.2
$(D_1), (D_2), (D_3)$ : 8.2.1
$D_n$ = dihedral group of order $2n$ : 8.5.5
$\partial$ = boundary : 8.4.4.3
$\varepsilon(x), \varepsilon_o(x)$ = error terms : 3.2.4
$\varepsilon_X, \varepsilon_X(p), \varepsilon_X(p^d)$ : 7.2.1
$\varepsilon$ : 8.5.6.1
$F$ = Frobenius endomorphism : 4.3
$<f,1>_G = \frac{1}{|G|}\sum_{g\in G} f(g)$ = mean value of $f$ on $G$ : 3.2.4
$f^N$ : 5.1.4; $\overline{f}$ : 7.1.1
$f(p), f_X(p), f^\lambda(p)$ : 8.1.1, 8.2.1
$f_*\mu$ (image of the measure $\mu$ by the map $f$) : 5.2.1
$\mathrm{Gal}(E/K)$ = Galois group of $E/K$ : 3.2.1
$\gamma$ : 8.2.3.4
$\Gamma_k = \mathrm{Gal}(k_s/k) = \mathrm{Aut}_k(\overline{k})$ : Conventions, 4.1

$\Gamma_S$ : 3.3.1, 6
$G_\ell, G_\ell^{\mathrm{zar}}$ : 8.3.2
$\mathbf{G}_m = \mathbf{GL}_1$ : 5.3.1
$g_p$ = geometric Frobenius : 6.1.1
**GSp** = symplectic similitudes : 8.5.5.
$h, h_X, h_i, h_{i,X}$ : 6.1.1
$\boldsymbol{h}$ : 8.3.4
$h(p, q, \lambda)$ : 8.2.3.2
$h_{X,\ell}, h_{i,X,\ell},\ h^i_{X,\ell},\ h^i_X(p^e)$ : 7.1.2
$h^w(p), h^w(p^e)$ : 8.1.1
$H^i(\overline{X}, \mathbf{Q}_\ell), H^i_c(\overline{X}, \mathbf{Q}_\ell)$ : 4.1
$H^i(X, \ell), H_i(X, \ell)$ : 6.1.1
$I_w, I_{\overline{v}}, I_{t'}$ = inertia group : 3.2.1, 4.8.2, 9.3.1.3
$\iota$ = embedding of a field into $\mathbf{C}$ : 4.5, 4.6, 5.3.1, 8.3.3
$\kappa(x), \kappa(v), \kappa(t)$ = residue fields : 1.2, 3.1.1, 9.1.1
$k_s, \overline{k}$ : Conventions
$K$ = ground field : 5.1
$K$ = residue field of the generic point : 9.1.2
$K$ = Sato-Tate group : 8.2.2
$K^0$ = identity component of the group $K$ : 8.2.3.1
$K_\sigma$ = connected component of the group $K$ : 8.4.3.1
$K[G]$ = group algebra of $G$ over $K$ : 5.1.1
$\mathrm{Li}(x), \mathrm{li}(x)$ : 3.1.2
$\lambda^k$ = $k$-th exterior power : 5.1.1.2
Max = maximal spectrum : 3.1.1, 9.3.1.1
mean$(f)$ : 3.3.3.5
$M(n), m(n,p), M'(n), m'(n,p)$ : 6.3.4
$m_\lambda$ : 8.4.1
$\mu$ = equidistribution measure : 8.1.2, 8.4.2.1
$\mu^{\mathrm{cont}}, \mu^{\mathrm{disc}}$ : 8.1.3.1, 8.4.2.5
$\mu_K$ = normalized Haar measure of $K$ : 8.4.2.1
$\mu_{\mathrm{Cl}}$ = image of $\mu_K$ under $K \to \mathrm{Cl}\, K$ : 8.2.2
$N : G \to \mathbf{C}^\times$ : 8.3.3
$N(p), N_f(p), N(p^e), N_f(\mathrm{mod}\, p^e)$ : 1.1
$NS$ = Néron-Severi group : 2.3.3, 8.5.6
$N_X(p^e),\ e \leqslant 0$ : 1.5
$N_X(p), N_X(q)$ : 1.2
$N_X(1), N_X(-1)$ : 6.1.2
$O(\ )$-notation : 1.3
$o(\ )$-notation : 3.1.3
$O_K$ = ring of integers of the number field $K$ : 3.1.1
$P$ = set of prime numbers : 3.4, 6
$\mathbf{P}_n$ = $n$-dimensional projective space : 2.1.1
$P_X, P_f$ : 7.2.4

$\pi_K(x), \pi(x)$ = counting function for primes : 3.1.1
$\pi_T(x), \pi_T^1(x), \pi_T^2(x)$ = counting functions for a scheme $T$ : 9.1
$\mathbf{Q}_\ell(1), \mathbf{Q}_\ell(-1), \mathbf{Q}_\ell(-d)$ : 4.5
$R_K(G)$, $R_K(G)^+$ : 5.1.1
$r_\lambda$ : 8.2.2
$\rho$ = Picard number : 8.5.6
$\rho$ = linear representation : 5.1, 5.3, 7.1.1
$\sigma, \sigma_q, \sigma_w, \sigma_{t'/t}$ = arithmetic Frobenius : 3.2.1, 4.4, 6.1, 9.3.1.3
$S$ = finite set of primes : 3.3, 6.1, 6.2, 6.3, 7.1, 7.2, 8.1, 8.2
$S_\ell = S \cup \{\ell\}$ : 4.8.2, 6.1.1, 6.1.2
$s_p$ : 8.2.2
$(ST_1), (ST_2), \ldots$ : 8.2
$\mathbf{SU}_2(\mathbf{C}) \stackrel{?}{=} \mathbf{SU}_2 \stackrel{?}{=} \mathbf{SU}_2(\mathbf{R})$ : 8.5.2.1
$|t|$ = number of elements of $\kappa(t)$ : 9.1.1
$\theta : K \otimes R_K(G) \to \mathrm{Cl}(G, K)$ : 5.1.1
$\underline{T}$ = set of closed points of $T$ : 7.3.1
$T, T', T_0, T_0'$ : 9.4
$\mathrm{Tr}(F)$ : 4.3
$\mathrm{Tr}_V$ : 5.1.1
$\mathbf{U}$ = unit circle in $\mathbf{C}^\times$ : 8.2.3.2
$\mathbf{USp}_4$ : 8.5.5
$[V]$ : 5.1.1
$V_K$ : 3.1.1
$V_{K,C}$ : 3.2.2
$V^N$ : 5.1.4
$v_p$ = $p$-adic valuation : 6.3.4
$V^{ss}$ = 5.1.1.1
$X_p$, $X_{/\mathbf{Q}}$ : 1.2
$X_0 = X_{/\mathbf{Q}}$ : 6.1.1
$\underline{X}$ = set of closed points of $X$ : 1.5
$X^\lambda$ : 8.2.1, 8.4.1
$Y_0 = Y_{/\mathbf{Q}}$ : 6.1.3
$s_p \in \mathrm{Cl}\, K$ : 8.2.2
$\chi$ = Euler-Poincaré characteristic : 1.4
$\chi_c$ = Euler-Poincaré characteristic with compact support : 1.4
$\chi_\ell$ = $\ell$-th cyclotomic character : 7.1.1
$<\chi, 1>_G = \frac{1}{|G|} \sum_{g \in G} \chi(g)$ = scalar product of $\chi$ and 1 : 3.3.3.5, 9.3.3
$\chi^N$ = character of $G/N$ : 9.3.3.
$\psi, \psi_\lambda$ : 8.4.2
$\Psi^e$, $\Psi^k$ = Adams operations : 3.3.2, 5.1.1.2
$w$ : 4.5, 7.1.1
$\boldsymbol{w}$ : 8.3.2, 8.3.3
$\omega$ : 8.2.3.3
$\ll$ notation : 3.1.2

# Index of Terms

affine-looking scheme : 7.2.5
almost equal (frobenian sets) : 3.3.2
analytic continuation principle : 5.2.2
arithmetically equivalent (number fields) : 6.1.3
arithmetic Frobenius : 3.2.1, 4.4
Artin's comparison theorem : 4.2
Betti numbers : 4.5
binary quadratic forms : 3.3.3.2
Brauer character : 5.1.1
Brauer-Witt theorem : 5.1.5 iv
Catalan numbers : 8.1.5.2
character of a $K$-linear representation : 5.1.1
Chebotarev density theorem : 3.2.3, 3.2.4, 6.2.1, 9.3
class function (on a group) : 5.1.1
clopen (subset) : 6.2.1
cohomological data (for Sato-Tate) : 8.2.1
compatibility : 7.1.3
computational problems : 2.1.4, 2.2.4
cubic surface : 2.3.3
decomposition subgroup : 3.2.1, 4.8.2, 9.3.1.3
degree of a closed point : 7.3.1.1
Deligne's theorems : 4.5
density : 3.1.3, 5.2.1, 9.2.1
Dwork's rationality theorem : 1.5
effective (element of a Grothendieck group) : 5.1.2
equidistribution : 3.2.2, 3.2.3, 8.1.2
Euler-Poincaré characteristic : 1.4
frobenian function, frobenian set : 3.3.2, 9.3.3
Frobenius automorphisms : 4.4, 6.2.1, 9.3.1.3
$\Gamma_{\mathbf{Q}}$-elementary : 5.1.5
generalized character : 5.1.1
geometric Frobenius : 4.4, 6.1
geometrically : 2.3.3
good reduction outside $S$ : 8.2.1
Grothendieck's theorem : 4.3

group algebra : 5.1.1
Haar density : 5.2.1.1
identity component (of an algebraic group) : 5.1.3
image (of a measure by a map) : 5.2.1
improved Deligne-Weil bound : 4.6
increasing function : 3.2.2
inertia subgroup : 3.2.1, 4.8.2, 9.3.1.3
integration over the fiber : 8.4.3.4
$\ell$-adic cohomology groups : 4.1
$\ell$-adic homology groups : 6.1.1
$\ell$-hyperelementary group : 5.1.5 iv)
$\ell'$-factor of an integer : 6.2.2
lattice (in a $\mathbf{Q}_\ell$-vector space) : 6.2.2
Laumon's theorem : 1.4
Lefschetz number : 4.3
$L$-function : 8.2.2, 9.3.3
Lie group over $K$ : 5.2.2
Lie groups data (for Sato-Tate) : 8.2.2, 8.2.3
$\ell$-hyperelementary (group) : 5.1.5
logarithmic integral : 3.1.2
lower density : 3.1.3, 9.2.1
mean value : 3.3.3.5
$m$-elementary (subset of $\mathbf{Z}^n$) : 6.2.2 Exerc.2
moments (of a measure) : 8.1.3.4
Hodge circle : 8.2.3.2
normalized Haar measure : 3.3.1, 5.2.1
positive (number) : Conventions
prime number theorem : 3.1.2, 7.3.1
P-problem : 2.1.4
proper support (cohomology) : 4.1
quarrable (subset) : 6.2.1, 8.1.2
representation ring of a group : 5.1.1
residually frobenian function : 3.3.3.4
Sato-Tate conjecture, Sato-Tate correspondence : 8.2, 9.5.4
semisimplification : 5.1.1
$S$-frobenian map, $S$-frobenian set : 3.3.1
size (of an Euler product) : 9.1.7
strong compatibility : 7.1.3
subquotient of a profinite group : 6.3.4
supernatural order of a profinite group : 6.2.2
support (of a measure) : 8.1.3.3
totally positive, totally real : 4.6.1
unitary analog : 5.3.1
unitary trick : 5.3
unramified : 9.3.1.3, 9.3.3, 9.5.3
upper density : 3.1.3, 9.2.1
value at 1 and at $-1$ of a frobenian function : 3.3.2

virtual Betti numbers : 7.1.3
virtual character : 5.1.1
weight : 4.5, 7.1.1
weight decomposition (of a virtual character) : 7.1.1
Weil conjecture, Weil integer : 4.5
Zariski density : 5.2.1.2
zeta function : 1.5

Printed in the United States
by Baker & Taylor Publisher Services